Intraepithelial Neoplasia

Intraepithelial Neoplasia

Edited by **Rob Ferrier**

New York

Published by Callisto Reference,
106 Park Avenue, Suite 200,
New York, NY 10016, USA
www.callistoreference.com

Intraepithelial Neoplasia
Edited by Rob Ferrier

International Standard Book Number: 978-1-63239-435-4 (Hardback)

Printed in the United States of America.

Contents

Preface

Written with the aim of serving experts to not only diagnose, but also understand the etiopathogenesis of precursor lesions, the book endeavors to identify its molecular and genetic techniques. This is a wide-ranging book devoted to preinvasive lesions of the human body. Chapters in this book consist of a significant quantity of novel information with an efficient catalogue list as well as the most recent WHO categorization of intraepithelial lesion such as intraepithelial neoplasia of the oral cavity, the eye, the breast and the uterus. This book has been modernized according to the most recent technological developments and can be described as brief, educational and practical text at all levels, discussing the important role of the molecular analysis of intraepithelial lesions.

This book is a comprehensive compilation of works of different researchers from varied parts of the world. It includes valuable experiences of the researchers with the sole objective of providing the readers (learners) with a proper knowledge of the concerned field. This book will be beneficial in evoking inspiration and enhancing the knowledge of the interested readers.

In the end, I would like to extend my heartiest thanks to the authors who worked with great determination on their chapters. I also appreciate the publisher's support in the course of the book. I would also like to deeply acknowledge my family who stood by me as a source of inspiration during the project.

Editor

Part 1

Intraepithelial Neoplasia of Cervix

Cervical Intraepithelial Neoplasia – Clinical and Etiological Aspects

Raghad Samir[1] and Dan Hellberg[2]
[1]Department of Obstetrics and Gynecology
Falun Hospital, Falun,
[2]Center for Clinical Research, Falun and
Department of Women's and Children's Health
Uppsala University, Uppsala,
Sweden

1. Introduction

Cervical cancer is one of the most common cancers among women worldwide. According to the most recent data, an estimated 466,000 new cases of cervical cancer occur among women worldwide each year, the vast majority of them in developing countries[1]. Overall global mortality rate is 60%, with large differences between industrialized countries and low-income countries. Of the 231,000 women who are deceased from cervical cancer annually, approximately 80 percent are from developing countries, where cervical cancer is the most common cause of cancer morbidity among women [2].

Organized screening programmes for cervical cancer using cervical cytology, papsmears, have been shown to be effective in decreasing mortality and incidence, and gynecological mass-screening has reduced the present incidence by more than half since the 1960s and 1970s in many high income countries [3-5].

Organized cervical cancer screening was implemented in Sweden in the mid 60s and since then a significant decrease in cervical cancer has been observed. Annually around 700,000 papsmears are taken in Sweden, with a total population of 9,000,000. Approximately 600,000 are taken within the mass-screening screening program, while the remaining is taken as part other gynaecological examinations. The overall incidence of cervical cancer declined by 67% over a 40-year period, from 20 Cases per 100 000 women (world standard rate) in 1965 to 6.6 per 100 000 women in 2005 [6]. During the last decade, however, the incidence has stabilized [7, 8]. Cancer of the ovaries, on the other hand, is decreasing by 1% to 2% annually based on data for the last 20 years. The increase of HPV infections are one cause that explain the present stability in the incidence of cervical cancer.

2. Background

Hippocrates in 450 b.c. was the first to describe cervical cancer as cancer of the uterus and added that it was a disease so destructive that it should better be left uncured than treated.

Aretaeus of Cappodocia 130-200 a.c. and in particular Aetius of Amida in 600 a.c. described uterine cancer as superficial and deep ulcers that would eventually infiltrate the uterus. He also described another type of cancer, which did not present with ulcers, but rather as a tumor in the uterus. In 1793, Mathew Baillie observed that cervical cancer did not cause enlargement of the uterus, but rather continuous ulceration of the uterus until the whole organ is destroyed [9].

In the early 19th, Rigoni-Stern, an Italian physician, examined the death records of the city of Verona between the years of 1760 and 1839. He noted that cancer of the cervix was common among married ladies and widows, but less among Jewish women, rare in unmarried ladies, and absent in nuns. This was the first report that incriminated reproductive and sexual events in the genesis of cervical neoplasms. Later, Rigoni-Stern found that the disease was very common among prostitutes, and cervical cancer thus became the poor and socially deprived women's disease [10]. von Scazoni, a German physician, reported in 1861 that female sexual activity outside marriage and sex not directed towards reproduction was the cause of cervical cancer development. Accordingly, woman who developed cervical cancer aroused suspicions of having engaged in 'too much sex' or having committed 'self-pollution'.

During the late 19th century, deaths due to cervical cancer in South Carolina, US, were observed to be much higher in black women, and socio-economic status was regarded as one of the risk factors. In 1895, John Clark examined 20 cases of cervical cancer treated by hysterectomy, Clark found that the disease in 15 hysterectomy specimens had extended past the margins of resection and he described the cervical tumor as 'peculiar cauliflower excrescence'.

Virchow stated in 1855 that 'every cell is derived from a cell' (*cellula a cellula*) and that human disease processes were essentially a disease of the cells. Virchow is considered the protagonist of the concept of *Zellular pathologie*, or pathology based on the study of cells. Virchows work can be considered a fore-runner to cervical cytological screening.

Background to cervical cytology screening

The papsmear (vaginal cytology) was developed by George Papanicolaou in the 1920s. Later, G. Papanicolou and the gynecologist Herbert Traut, in 1941 published the first major article on the use of vaginal smears for the diagnosis of cancer of the uterus. Soon thereafter, the papsmear (named after Papanicolou), was born and it is still one of the most sensitive, simple and effective cancer screening tests.

Simultaneously, Hans Hinselman and Leitz technicians devised the first working binocular colposcope. In 1925, he published the first paper on colposcopy, and later on in 1933 the book 'Einfurthrung in die Kolposcopi' was published. Colposcopy was developed furthermore in 1925 by Hinselman and Eduardwirth, but routine colposcopic examinations were confined to Germany until the 1960s. In the United States, as early as 1929, Levy described the importance to study the genital tract with some degree of magnification and subsequently Emmert published an article introducing the colposcope to North American physicians. By 1932 the colposcopic technique was used in a few centers. The present form of colposcopy started in 1953 when Bolten introduced the modern colposcope in United States. Initially, it served as a tool to identify women with asymptomatic early invasive disease. Subsequently, it has also helped physicians identify preinvasive squamous neoplasia of the cervix [11]. At a meeting of the American College of Obstetricians and Gynecologists in Miami in 1964, a group of enthusiastic colposcopists identified the need for a colposcopy society. Thereafter, through the dedicated efforts of many members of the society, various colposcopy courses were initiated. In 1981, Hamou introduced the microhystreoscope for the examination of the cervix and endocervical canal. This provided a panoramic and contact microscopical observation of stained cells in vivo at high magnification.

Treatment: Historical background

The first radical hysterectomy was described by John G. Clark at the Johns Hopkins Hospital, US in 1895. At a pathological examination of 20 cases treated by hysterectomy, Clark found that the disease had extended past the margins of resection in 15 cases. Influenced by the surgical doctrines of William Halsted, he developed an operative technique that is today recognized as the first true radical hysterectomy [12]. The operation was modified and popularized by Ernst Wertheim, whose experience was impressive in magnitude, follow-up, and descriptions of complications associated with the procedure. In 1898, Wertheim introduced abdominal radical hysterectomy with the removal of the adjacent medial portion of parametria and the upper part of the vagina. In 1945, pelvic lymphadenectomy was added to radical abdominal hysterectomy and its gained the name modified Wertheim. [13].

Screening program in Sweden

Cervical cancer screening has been linked to population- and pathological registers in specific counties since the beginning of the 1960s, and all Swedish counties had computerized screening programs in 1993. The screening is free of charge. All women aged 23-50 are invited every third year for gynecological screening. Women between 51-60 years of age are invited at 5 years intervals. The papsmear is taken by midwives and all results are reported to the patient. If the smear shows abnormal findings or CIN, the patient will be referred to a gynecologist for colposcopical examination. HPV DNA-testing is used with triage of ASCUS and CIN1 and in the follow up of CIN2 and CIN3 (see below).

Histopathology

The most common type of the cervical cancer is squamous cell carcinomas and these represent 80% of all cervical cancer, while 15-20% are adenocarcinomas.

Dysplasia is the premalignant squamous cell abnormalities that range from mild, moderate and severe dysplasia, and eventually carcinoma in situ, but this classification has been replaced by cervical intraepithelial neoplasia (CIN). CIN is also used for histological abnormalities that are histopathologically diagnosed in cervical biopsies.

CIN1 (mild dysplasia) is a low grade lesion with atypical cells in the basal, lower third, of the epithelium. Viral cytopathic effects by HPV (koilocytotic atypia) are often present. Another name is low-grade SIL (squamous epithelial lesion).

CIN 2 (moderate dysplasia) is also called high grade lesion HSIL. It refers to moderately atypical cellular changes confined to the basal two-thirds of the epithelium, with preservation of epithelial maturation in the superficial parts of the epithelium.

CIN 3 (severe dysplasia and cancer in situ), also HSIL, refers to severely atypical cellular abnormalities encompassing more than two-thirds or the complete epithelium.

Prognosis

Most dysplasias remain stationary or regress, but some dysplasias progress to carcinoma in situ and subsequently to invasive cancer. The progress of HPV infection/CIN1 to CIN3 is estimated to 10 years, but progression in 1-3 years is not uncommon. Similar estimates are considered for the progress of carcinoma in situ to invasive cancer. The progression rate of CIN 3 to invasive squamous cell cancer has been reported to be 12-30% in different investigations, and might depend on the size of lesion, the age of patient, immunological factors and the characteristics of the study population.

Management of the abnormal papsmear

Management differs between countries, hospital and economical resources. Liquid cytology is increasingly used instead of papsmear, as cells could be spared for HPV diagnosis. A colposcopically directed cervical biopsy undergoes histopathological examination. When the microscopical examination is normal, repeated papsmear shall be performed within 12 months, and when negative, no further controls are required. When the papsmear continuous to be abnormal, a new colpscopical examination is required. When the initial HPV-DNA test was positive, if evaluated, but with no cytological CIN is found, a directed colposcopical biopsy shall be performed and further management depends on the results.

An alternative strategy has been suggested, i.e. that all women above 35 years of age with cytological ASCUS (atypical cells of undermined significance) or CIN1 shall be examined for detection of HPV-DNA, as this test has a high sensitivity for CIN 2 and CIN3 compared to repeated papsmears.

When biopsy has been taken with directed colposcopy without HPV testing, the approach according to this alternative management will include colposcopy and a new cytological test (PAP) within 6 months in women with ASCUS or CIN1.

See-and-treat

See-and-treat is based on colposcopical findings without previous histopathological grading. This includes an electrosurgical loop excision of the cervical transformation zone (ELECTZ), a superficial cone biopsy, that facilitates the subsequent histological diagnosis and might be the sole treatment of CIN if microscopically radical, thus eliminating the need for a preliminary cervical biopsy and additional patient visits. Requirements for the procedure are an abnormal cervical papsmear and a colposcopical suspicion of CIN. This procedure is not recommended in case of ASCUS and CIN1, as in general there is spontaneous regression of mild lesions. See-and-treat could also be used when the colposcopy findings after an abnormal papsmear are not conclusive.

In CIN2 and CIN3 colposcopically directed examinations including the vagina are essential. When indicated, a biopsy should be taken. If no lesion can be detected, papsmear shall be taken and if it positive, a diagnostic conisation is preferred. When the biopsy shows CIN2 or CIN3 a therapeutic conisation should be performed.

Management of glandular cell atypia or AIS (adenocarcinoma in situ) is similar to that of squamous cell high grade lesions, but should include endometrial biopsy in women above 40 years of age. The conisation should be extended higher in the cervical canal than in squamous cell lesions.

Treatment methods

All methods of treatment of cervical intraepithelial neoplasia are surgical, and might differ with histological findings, extension of lesion, adverse reactions and cost effectiveness [14]the age of the patient, the possibility of pregnancy, as extensive treatment methods can decrease fertility and pregnancy outcome. Treatment methods are ablative or excisions.

Ablative methods

Cryotherapy: By using liquid nitrogen at -270 C° in a closed cylinder with cervical application for 5-10 minutes the abnormal tissue are frozen to temperatures as low as -29, and the tissue will be removed.

Laser ablation: A laser beam is used to evaporate the abnormal tissue with the aid of application of acetic acid and colposcopy in order to visualize the affected area in the cervix.

Excisional

Cold knife, using a scalpel, or laser conization: The affected area and the complete transformation zone are removed with a cone biopsy, where the size depends on the lesion, by traditional surgery.

Loop electrosurgical excision [LEEP, LLETZ]): This is a surgical procedure that uses an electrified wire to remove tissue from the cervix. It is done under local anaesthesia and colposcopy examination.

There were no differences in the effectiveness and outcome between ablative or excisional surgery according to a systematic review of randomized controlled trials in women who underwent treatment of low- and high-grade CIN with cryotherapy, laser ablation, or LEEP 15. Cold knife or laser conization is by many investigators considered as the most effective treatments.

Topical treatment with Acyclovir (nucleotide analogue), cyclooxygenase inhibitors and other pharmaceutical treatments have been suggested and tried. The success rates have been poor.

Vaccination

Prevention by vaccination will probably decrease the incidence of HPV infections, CIN and cervical cancer in the future. The identification of the HPV oncogenes E6 and E7 led to the development of effective vaccines with immunological activation of HPV antibodies, but is at present only directed against HPV 16 and HPV 18. Some cross reactions with other high risk HPV types have, however, been observed against other high risk HPV types [16]. As 13-18 HPV types are considered high-risk, conclusive results will not be available in 10-20 years.

Clinical studies have demonstrated that HPV 16 L1 VLP vaccines are well tolerated and generate high level of antibodies against HPV 16. The prophylactic polyvalent vaccine against oncogenic HPVs in young girls prior to the onset of sexual activity is the key for prevention and avoidance of the disease. An early double-blind, randomized study showed that the vaccine was well tolerated and with high immunological response [17]. In another study 2392 young women 16-23 years of age with negative HPV-16 DNA were included. They were randomly assigned to receive the HPV-16LVLP vaccine or placebo at day 0, and at 2, and 6 months. Forty-one women (3.8 per 100 woman years) in the placebo group acquired HPV-16 infections, including 9 cases of CIN, during follow-up, while no woman who received the HPV-16 vaccine, developed infection or CIN. In one study, HPV types 16, 18, 42, 31, 33, 52 and 35 were in descending order of frequency the most common types in cervical cancer.[18]

3. Risk factors for cervical neoplasia

Among risk factors for CIN, human papillomavirus infection, smoking and sex steroid hormones, in general hormonal contraceptives are the most studied and important. In countries and areas associated with many childbirths, like in developing countries or where the use of contraceptives because of religion, parity is still a major risk factor

Human papillomavirus infections

HPV infections are the most common genital infection worldwide. It is sexually transmitted and mostly clinically silent and self-limiting. Some women remain persistent carriers of the viral infection and become at high risk of progression to CIN and invasive cervical cancer [19].

The lifetime risk of genital HPV infection is approximately 80%. For many HPV infected women (80%), the immune defence will eliminate the infection in general after approximately 12 months. In the remaining cases, progression to cytological abnormalities and CIN is observed in 5% to 10% of persistent HPV infections. After an interval of 7-15 years less than 1% of these infections lead to carcinoma.

Human papillomavirus belongs to the family Papillomaviridae, which includes viruses infecting many different vertebrates. They are strictly species- and organ-specific. HPV are small DNA viruses. The relatively small viral genome of 8000 base pairs are organized in three different regions: the long control region, also called the non-coding region, and the two coding regions, the late (L) and the early (E) regions. The early coding region in human HPV types is divided into E1, E2, E4, E5, E6 and E7. and encodes for the proteins with different regulatory functions [20].

Different HPV types exhibit a type-specific tropism either for squamous epithelium or mucosal sites. Viremia or systemic infections are absent. Despite low or undetectable antibody levels following infection, It is unknown if the HPV type-specific immunity is [21] is lifelong.

The HPV life cycle in the cervix is confined to the epithelium. The border between squamous cell epithelium covering the vagina and glandular epithelium covering the uterus, the transformation zone is the target for HPV invasion. Most HPV-related lesions resolve spontaneously, and progression to cervical neoplasia is a relatively rare event. A key factor in allowing disease to progress is the ability of HPV to evade the immune system and establish a persistent infection. Approximately 50% of infected individuals fail to demonstrate or produce a detectable antibody response to HPV. In those who respond there is no full protection from future HPV infections.

High risk HPV infections

Among more than approximately 150 HPV types, 13 HPV types are considered high-risk and 5 HPV types as moderate-risk for cervical neoplasia. Globally, HPV 16 and HPV 18 are the predominant oncogenic types, accounting for more than 70% of all cervical infections [18, 22]. Low-risk types are rarely found in cervical neoplasias, but some types, in particular HPV 6 and 11 are associated to benign genital condylomas.

High-risk HPV DNA is according to our own studies, detected in 37% in low-grade lesions (CIN1 and borderline lesions), 89% in high-grade lesions (CIN2 and CIN3), and in 40% of cytological CIN in papsmears, but with no information of histological CIN grade [23]. In invasive cancer HPV DNA are detected in close to 100%. This indicates that while HPV is a necessary factor for cervical cancer, other factors could be responsible for development of CIN.

There are several steps in the pathway from HPV infection to CIN and cervical cancer. The initial viral entry is the target cells of the basal epithelium. HPV DNA integrates into the host genome and the HPV oncogenes E6 and E7 are expressed in some cases of CIN2 and in most cases of CIN3 (carcinoma in situ). The results of integration are cytogenetic instability and uncontrolled cell growth (immortalization). For malignant transformation, CIN3 or invasive cancer, HPV DNA integration is necessary. HPV itself may transform cervical cells from normal into CIN3, but is not sufficient in developing CIN3 into invasive cancer. Co-factors are needed and will be discussed below.

Integration of the HPV genome into the host genome frequently leads to disruption of the E2 gene that regulates the expression of the two major oncogenes, E6 and E7. Protein products of these oncogenes are responsible for transforming and immortalizing cells, which may lead to CIN3 or invasive cervical cancer. The viral oncoproteins E6 and E7 degrade two key tumor suppressors, p53 and retinoblastoma protein, respectively [24]. p53 and retinoblastoma proteins cause cell cycle arrest, allowing for repair of mutant DNA or inducing apoptosis, programmed cell death. Inactivation leads to unscheduled progression through the cell cycle and proliferation, which is required for development and maintenance of malignant cells.

4. Smoking

Epidemiological studies

During the 1980s, studies started to appear, where the correlation between smoking and cervical neoplasia, independent of sexual risk factors was evaluated. As far back as 1966 it was reported that smoking was twice as common among women with CIN as those without, but it was regarded as a confounder for sexual risk behaviour [25].

A number of studies from 1980 and onwards confirmed that smoking was an risk factor, independent of sexual risk behaviour, for CIN [26]. The first report on smoking and CIN that adjusted for sexual risk behaviour on CIN estimated relative risks for smokers to be above 2.0 in multivariate analyses, and slightly, but not substantially, higher in crude analyses. Thus, the association between smoking and sexual risk behaviour was not very strong [27].

In 1983 followed three studies on CIN, one by us and another two independent studies (Hellberg, Valentin et al. 1983; Lyon, Gardner et al. 1983; Trevathan, Layde et al. 1983). They all confirmed the results of the initial study. We suggested, and later performed, studies on cervical mucus in smokers. There was a slight decrease in odds ratios between crude and multivariate analyses in these studies. Independent odds ratios in all three studies were between 2.0 and 3.0. In addition, our results indicated that passive smoking could play a role. Many confirmatory studies where proper adjustments were made have followed these four initial studies smoking habits might ever be more important in CIN than in invasive cancer. Thus, there are reports of a higher relative risk with CIN compared to invasive cancer, and smoking. It suggests that smoking is particularly involved in early carcinogenesis and might be a biological co-factor to a progressive HPV infection[28]. In some studies, women above 40 years of age show lower odds ratios for smoking and CIN compared to younger women [29, 30]. The role of the hormonal environment in premenopausal women will be discussed below.

One study [30] also found a strong trend with increased risk by pack-years of smoking and age of starting smoking. A number of early, unadjusted studies were also able to show a dose-response relationship, a prerequisite in most epidemiological studies where exposure is analysed. A number of later studies have confirmed the dose-response relationship for smoking [31]. Some studies on previous smokers have reported substantially decreased risk to acquire CIN after some years of smoking cessation [31-33].

Decreasing odds ratios in multivariate analyses compared to crude analyses are of some concern. There is always a possibility of residual confounding, i.e. variables that was not controlled for. Age at first intercourse and number of lifetime sexual partners may not entirely reflect sexual risk behaviour. Factors such as sex at first date, sex with casual and unknown partners, sex tourism and anal sex all increase the risk for acquiring a sexually transmitted infection. Moreover, the male's sexual risk behaviour is rarely, if ever, controlled for. As in most epidemiological studies they must be confirmed by similar studies *and* additional biological and experimental evidence is necessary to support the results, to be considered conclusive.

An interesting finding in one study was that smoking attributed little to the risk in women with many sexual partners, but was an important risk factor in women who had only had 0-1 sexual partners and also with increasing risks by years of smoking. The relative risk was impressingly 3.7 in women who had smoked for more than 20 years. The results adds to the findings that smoking also has a role independent of HPV, as these women could not be supposed to practising a sexual risk behaviour [34].

In a calculation of attributable risk (PAR) for CIN for a large number of risk factors, smoking (29%) was second to number of sexual partners (57%), while attributable risk for long-term oral contraceptive use was only 8%. [35].

Finally, a number of meta-analyses on the role of smoking in CIN and cervical cancer have been performed [36-39]. All meta-analyses concluded that smoking epidemiologically seemed to be an independent risk factor for cervical cancer. There were some indications that the risk was higher in HPV-infected compared to non-infected women. Calculated pooled odds ratios ranged from 1.95 to 2.17.

Smoking and human papillomavirus infections

It has been suggested to evaluate the importance of smoking only in women with negative HPV status. HPV detection in both CIN III and invasive cervical cancer is approaching 100% why such studies would be hard to conduct. The presence of HPV is lower in low-grade CIN, and searching for a correlation to smoking might give new information. An epidemiological association between HPV infection and smoking has been searched for in numerous studies. In most studies there are significant corrrelations between smoking habits and HPV infection. It has been speculated that smoking-induced impaired immune defence is behind the correlation. Declining odds ratios after adjustments raise suspicions of residual confounding, i.e. risk behaviours or factors that were not adjusted for. [26].

HPV, smoking and cervical neoplasias

More important, also clinically, than a possible association between a cervical HPV infection with normal epithelium and smoking, would be if smoking in addition to HPV is also involved in the transformation from normal to cervical neoplasia. One approach is to study if smoking is more prevalent in HPV infected women with high-grade CIN compared to HPV infected women with low-grade CIN. Indeed, some studies that found that there was a higher (4.4) correlation between smoking and CIN II-III than to CIN I. [40]. Smoking frequency was reported to be increased from 16% in women having no CIN or CIN I, to 41% in those who had CIN III [41]. There are also studies with discrepant results.

Of interest in this review are some studies that investigated the association between CIN and smoking after adjustment for current HPV infection and this was claimed to insignificantly decrease the correlation between CIN II-III and smoking habits. With presence of HPV infection odds ratios (3.0) were unchanged after adjustments which is somewhat surprising. In studies where HPV negative and positive women were analysed separately, odds ratio for HPV negative women was still significantly increased in smokers [42]. Similar results have been reported in other studies. If these results are true carcinoma in situ (CIN III) can develop in smokers, maybe in combinations with other potential carcinogens, even without a current HPV infection.

Experimental studies

As stated above, experimental results must be added to epidemiological findings. The first biological explanation in humans was our finding of nicotine levels that were 40 times higher in cervical mucus, directly collected from the cervical canal with a syringe, compared to serum levels in healthy smokers. In addition, the stable nicotine metabolite cotinine were found in almost four times higher concentrations in mucus than in serum [43]. In a larger study on smoking women with current CIN we could confirm the results. While nicotine and cotinine could not be measured in serum of non-smokers and passive smokers both substances were found in small amounts in cervical mucus. There was a dose-response relationship by smoking intensity and nicotine/cotinine levels in the cervix [44]. In addition, we measured these tobacco constituents in another female genital gland, the follicle fluid of the ovaries. Nicotine and cotinine levels were found to be equal to those in serum [45]. Subsequent studies confirmed the results of the finding of tobacco constituents in the cervix [46, 47]. A problem was that these studies used a cervical flush technique and direct nicotine levels in mucus could not be estimated.

It is unclear whether nicotine and cotinine in itself exert carcinogenic effects. In cell lines derived from human normal, HPV transfected cells nicotine was added to tissue culture plates at concentrations we found in cervical mucus and in higher concentrations. Nicotine enhanced cell proliferation in all three lines [48].

We tried to analyse carcinogenic tobacco products, during the mid-eighties there was not enough sensitive methods to detect tobacco carcinogens, in particular tobacco-specific nitrosamines and polynuclear aromatic hydrocarbons (PAH), i.e. benzpyrenes. During the late 1990s, the presence of the highly carcinogenic, tobacco-specific nitrosamine NNK and benzo(a)pyrene in smoker's cervical mucus were finally found by our previous collaborating laboratory (Prokopczyk, Cox et al. 1997; Melikian, Sun et al. 1999).

During the 1950s and the 1960s a number of animal studies aimed at transforming normal cervical epithelium to malignant. Coal-tar PAHs, like benzpyrenes was reported to induce invasive squamous cell carcinoma [49], but studies of human cervical cell lines have been difficult to interpret [50, 51].

Smoke condensate – in vitro studies

Smoke condensate has been administered to HPV immortalized cell lines. When two human HPV 16 containing cell lines underwent condensate treatment for each passage up to 26 months, they progressed to malignant tumours with few exceptions. Treatment with smoke condensate, but not without, formed invasive cervical cancer when injected in nude mice [52]. The same research group reported that in a HPV 18 immortalized cell line from the transformation zone, that addition of smoke condensate, but not without, was followed by malignant transformation [53].

Tumor markers – in vivo studies

There are still few cervical tissue studies specifically investigating smoking and the levels of proteins considered to be tumor markers in invasive cancer, and in particular in CIN. We studied 17 tumor markers in CIN and normal epithelium, correlated to smoking habits. Some of the tumor markers showed no or entire expression, but most were possible to evaluate. In normal epithelium there were no correlations to expression of tumor markers. In CIN, the tumor suppressors p53 and FHIT, and the immunologic marker IL-10 were underexpressed, while the proliferation marker Ki-67, and Cox-2, involved in many carcinogenic processed, were overexpressed. Thus, this provides in vivo biological evidence for a direct promoter role of smoking in CIN [54].

In another study of women with normal histology, or CIN I to CIN III, smoking was significantly associated to CIN, while they could not confirm our results of smoking and Ki-67 expression. Odds ratios were, however, 2.0-4.2 depending on smoking intensity which indicates that the study did not have enough power. Trend for number of cigarettes per day was of borderline significance [55].

We also studied expression of 14 tumor markers and correlation to smoking in invasive cervical cancer tissue. Many of these markers were also included in the study of CIN. Interestingly, only decreased p53 and increased Cox-2 expression were significantly correlated to smoking both in CIN and cancer. In cancer, also decreased expression of the tumor suppressor LRIG1, and increased expressions of the angiogenesis protein VEGF correlated to smoking, in contrast to CIN [54, 56, 57]. This indicates that the biological roles of smoking might not be entirely similar in CIN and invasive cancer.

Sex steroids – hormonal contraceptives

Multiparity is a classic risk factor for CIN and cancer, but might be a confounder for high-risk sexual behavior, in particular before the introduction of commonly used modern contraceptives, but that is still not the situation in many parts of the world. It still gave an indication that reproductive factors were involved in the etiology of CIN. Most of the major studies restricting the analysis to HPV-positive women, also report an increased risk for cervical neoplasia with increasing parity. It has been suggested that the increased exposure of the cervical transformation zone, where cervical neoplasia is initiated, after pregnancy might facilitate HPV infection [58]. In that case, the hormonal influence would only be secondary.

Oral contraceptive (OC) use early emerged as an epidemiological risk factor for cervical neoplasia, but only in the early 1980s, studies with adjustment for other risk factors, i.e. sexual risk behavior and smoking, appeared [27, 59]. It might be expected that women with oral contraceptive use are more sexually active than women without. Sexual abstinence, marriages or other characteristics will decrease the necessity of contraceptive use. Parity could be lower among oral contraceptive users and could introduce a negative bias. Smoking habits might correlate to sexual risk behavior, and detection bias due to frequent papsmear evaluations, must be taken into account. Thus, there are numerous pitfalls in epidemiological studies.

Oral contraceptives increase serum levels of sex steroid hormones. In epidemiological studies it is not possible, to confirm the theory of sex steroid hormones as causal co-factors to HPV in the transition of normal epithelium to CIN and invasive cancer. When the first epidemiological studies with adjustment for sexual risk behavior were published, that also included smoking habits [59], it became clear that OC use, but only long-term use, i.e. more than 4-5 years, was independently correlated to cervical neoplasia, irrespective of sexual risk behavior and smoking [27, 59]. Odds ratios were in general moderate, 1.5-2.0, but significant. Discrepant results were found in some subsequent studies, but in those cases commonly a tendency of a correlation was observed.

OC use must be investigated for a possible correlation to HPV infection, to exclude that OCs are merely bystanders to HPV. If OC use is a risk factor independent of HPV infection, it strengthens the evidence that OCs are true biological co-factors. Several studies have investigated an eventual correlation to HPV infection and adjusted the results for sexual risk behavior [23, 60-62]. These studies found independent correlations between OC use and cervical neoplasia. Interestingly, we found that use of high dose OCs, but not low dose OCs, was significantly associated with HPV infection [60]. Available studies does not indicate that any hormonal contraceptive influence prognosis in CIN or invasive cancer.

Immunity

Sex steroid hormones exert effects on immune responses. Overall, progesterone is associated with tumor suppression, allowing for immunological escape for HPV infected cells. Estradiol seems to be associated with an increased immune defense. During pregnancy, the natural killer cell activity is suppressed, indicating a decreased immunological response. The clinical role of such findings is unclear.

In an early study, progesterone and glucocorticoid response elements were identified in the long region of several types of the HPV genome, and administration of progestins increased expression of the oncogenes E6 and E7, considered crucial in cell transformation [63] . In another study on HPV positive cell lines, progesterone treatment enhanced the colony formation, while no effect was observed on HPV negative cell lines [64]. These studies on human cell lines support the notion that progesterone is the major sex steroid co-factor in cervical cancer. It was, however, also reported that estrogen treatment stimulated HPV 16 transcripts in another cell line, while progesterone did not [65]. Finally, p53 expression was increased in HPV cervical cancer cell lines after treatment with high doses of estradiol, a favorable effect, but not with low or medium doses, a possible favorable effect in tumor suppression [66].

Serum levels of progesterone and estradiol

The idea of studying cervical neoplasms by correlating clinical variables to serum hormone levels is attracting as it reflects physiologic conditions. We performed two studies on premenopausal women with cancer. Outcome in the first experimental study was S-phase fraction in tissue, a marker of proliferation. High serum progesterone, but not estradiol, levels directly correlated to a high S-phase fraction and after adjustment for eight variables, only serum progesterone and smoking emerged significantly correlated to proliferation [67].

In a clinical study, mortality was studied and adjustment was made for a number of variables, e.g. clinical cancer stage, the most important variable for prognosis. Premenopausal women, who eventually died from cervical cancer, with high serum estradiol showed increased survival-months, compared to those with low serum estradiol, while women with high progesterone levels showed decreased survival-months than those with low levels. A estradiol-progesterone ratio was constructed and the combination of high estradiol and low progesterone correlated significantly to a longer survival, and vice verse [68].

We performed two studies, one clinical and one laboratory, where the above criteria were taken care of. In both studies, all women were analysed together, and pre- and postmenopausal women were also analysed separately. Analyses in the former group showed no differences regarding the variables included, and results were presented only for the premenopausal group. Outcome was S-phase fraction, i.e. the percentage of dividing cells in the cancer tissue that is a marker of proliferation and cancer growth. Close to all tumors where serum progesterone was high, had a high S-phase fraction. There were no correlations to serum estradiol levels and after adjustment for eight variables, only serum progesterone and smoking emerged significantly correlated to proliferation. This supports the theory that progestins are promoters of cancer growth and correlated to poor prognosis [67].

In a clinical study, women with high serum estradiol, but who eventually died from the disease, however, showed increased survival-months, compared to those with low serum estradiol, but this was nonsignificant. Women with high progesterone levels, on the other hand, showed decreased survival-months than those with low levels. A estradiol-progesterone ratio was constructed and the combination of high estradiol and low progesterone increased the prognosis prediction [68].

To find evidence that estradiol or progesterone correlates to CIN grade a study where both cases and controls were HPV positive was conducted. None of the hormones measured did correlate to lesions more or equal to CIN2 compared to low-grade CIN. Serum levels of sex steroids thus have not proved to be useful in distinguishing low- and high-grade lesions [69].

We evaluated tumor marker expression and serum progesterone levels in CIN, and found a significantly higher expression of Cox-2, low retinoblastoma protein (tumor suppressor) and low p16 (tumor suppressor) expression with high progesterone levels, the former were independent of CIN grade. No correlations between serum estradiol and tumor marker expressions were found. It could be concluded that progesterone levels CIN are associated with a negative tumor marker pattern, as was the case in oral contraceptive users [70]. In summary, available results indicate that oral contraceptives and high serum progesterone levels exert unfavorable effects in CIN, both epidemiologically and in laboratory studies, while the role of estrogens are unclear [70].

5. References

[1] Gustafsson L, Ponten J, Bergstrom R, Adami HO. International incidence rates of invasive cervical cancer before cytological screening. Int J Cancer 1997;71:159-65.

[2] Sankaranarayanan R, Ferlay J. Worldwide burden of gynaecological cancer: the size of the problem. Best Pract Res Clin Obstet Gynaecol 2006;20:207-25.

[3] Laara E, Day NE, Hakama M. Trends in mortality from cervical cancer in the Nordic countries: association with organised screening programmes. Lancet 1987;1:1247-9.

[4] Hakama M, Rasanen-Virtanen U. Effect of a mass screening program on the risk of cervical cancer. Am J Epidemiol 1976;103:512-7.

[5] Fidler HK, Boyes DA, Worth AJ. Cervical cancer detection in British Columbia. A progress report. J Obstet Gynaecol Br Commonw 1968;75:392-404.

[6] Andrae B, Kemetli L, Sparen P, et al. Screening-preventable cervical cancer risks: evidence from a nationwide audit in Sweden. J Natl Cancer Inst 2008;100:622-9.

[7] National Board of Health and Welfare. Cancer incidence in Sweden. Official statistic of Sweden 2008.

[8] Cancer incidence in Sweden. National Board of Health and Welfare. Official statistic of sweden 2008.

[9] Gasparini R, Panatto D. Cervical cancer: from Hippocrates through Rigoni-Stern to zur Hausen. Vaccine 2009;27 Suppl 1:A4-5.

[10] De Palo G. Cervical precancer and cancer, past, present and future. Eur J Gynaecol Oncol 2004;25:269-78.

[11] Nyberg R, Tornberg B, Westin B. Colposcopy and Schiller's iodine test as an aid in the diagnosis of malignant and premalignant lesions of the squamous epithelium of the cervix uteri. Acta Obstet Gynecol Scand 1960;39:540-56.

[12] Grigsby PW, Herzog TJ. Current management of patients with invasive cervical carcinoma. Clin Obstet Gynecol 2001;44:531-7.

[13] Gaffney DK, Erickson-Wittmann BA, Jhingran A, et al. ACR Appropriateness Criteria(R) on Advanced Cervical Cancer Expert Panel on Radiation Oncology-Gynecology. Int J Radiat Oncol Biol Phys 2011.

[14] Salani R, Backes FJ, Fung Kee Fung M, et al. Posttreatment surveillance and diagnosis of recurrence in women with gynecologic malignancies: Society of Gynecologic Oncologists recommendations. Am J Obstet Gynecol 2011;204:466-78.

[15] Nuovo J, Melnikow J, Willan AR, Chan BK. Treatment outcomes for squamous intraepithelial lesions. Int J Gynaecol Obstet 2000;68:25-33.

[16] Borysiewicz LK, Fiander A, Nimako M, et al. A recombinant vaccinia virus encoding human papillomavirus types 16 and 18, E6 and E7 proteins as immunotherapy for cervical cancer. Lancet 1996;347:1523-7.

[17] Koutsky LA, Ault KA, Wheeler CM, et al. A controlled trial of a human papillomavirus type 16 vaccine. N Engl J Med 2002;347:1645-51.

[18] Munoz N, Bosch FX, de Sanjose S, et al. Epidemiologic classification of human papillomavirus types associated with cervical cancer. N Engl J Med 2003;348:518-27.

[19] Trottier H, Franco EL. Human papillomavirus and cervical cancer: burden of illness and basis for prevention. Am J Manag Care 2006;12:S462-72.

[20] Syrjanen SM, Syrjanen KJ. New concepts on the role of human papillomavirus in cell cycle regulation. Ann Med 1999;31:175-87.

[21] Viscidi RP, Snyder B, Cu-Uvin S, et al. Human papillomavirus capsid antibody response to natural infection and risk of subsequent HPV infection in HIV-positive and HIV-negative women. Cancer Epidemiol Biomarkers Prev 2005;14:283-8.

[22] Smith JS, Lindsay L, Hoots B, et al. Human papillomavirus type distribution in invasive cervical cancer and high-grade cervical lesions: a meta-analysis update. Int J Cancer 2007;121:621-32.

[23] Samir R, Asplund A, Tot T, Pekar G, Hellberg D. High-Risk HPV Infection and CIN Grade Correlates to the Expression of c-myc, CD4+, FHIT, E-cadherin, Ki-67, and p16INK4a. J Low Genit Tract Dis 2011.

[24] Munger K, Baldwin A, Edwards KM, et al. Mechanisms of human papillomavirus-induced oncogenesis. J Virol 2004;78:11451-60.

[25] Naguib SM, Lundin FE, Jr., Davis HJ. Relation of various epidemiologic factors to cervical cancer as determined by a screening program. Obstet Gynecol 1966;28:451-9.

[26] Hellberg D. Smoking and cervical cancer. in Research focus on smoking and women´s health, eds Tolson, KP and Veksler EB 2008:19-60.

[27] Harris RW, Brinton LA, Cowdell RH, et al. Characteristics of women with dysplasia or carcinoma in situ of the cervix uteri. Br J Cancer 1980;42:359-69.

[28] La Vecchia C, Franceschi S, Decarli A, Fasoli M, Gentile A, Tognoni G. Cigarette smoking and the risk of cervical neoplasia. Am J Epidemiol 1986;123:22-9.

[29] Lyon JL, Gardner JW, West DW, Stanish WM, Hebertson RM. Smoking and carcinoma in situ of the uterine cervix. Am J Public Health 1983;73:558-62.

[30] Trevathan E, Layde P, Webster LA, Adams JB, Benigno BB, Ory H. Cigarette smoking and dysplasia and carcinoma in situ of the uterine cervix. JAMA 1983;250:499-502.

[31] Ylitalo N, Sorensen P, Josefsson A, et al. Smoking and oral contraceptives as risk factors for cervical carcinoma in situ. Int J Cancer 1999;81:357-65.

[32] Brinton LA, Schairer C, Haenszel W, et al. Cigarette smoking and invasive cervical cancer. JAMA 1986;255:3265-9.

[33] Kjaer SK, Engholm G, Dahl C, Bock JE. Case-control study of risk factors for cervical squamous cell neoplasia in Denmark. IV: role of smoking habits. Eur J Cancer Prev 1996;5:359-65.

[34] Nischan P, Ebeling K, Schindler C. Smoking and invasive cervical cancer risk. Results from a case-control study. Am J Epidemiol 1988;128:74-7.

[35] de Vet HC, Sturmans F. Risk factors for cervical dysplasia: implications for prevention. Public Health 1994;108:241-9.

[36] Appleby P, Beral V, Berrington de Gonzalez A, et al. Carcinoma of the cervix and tobacco smoking: collaborative reanalysis of individual data on 13,541 women with carcinoma of the cervix and 23,017 women without carcinoma of the cervix from 23 epidemiological studies. Int J Cancer 2006;118:1481-95.

[37] Franceschi S. The IARC commitment to cancer prevention: the example of papillomavirus and cervical cancer. Recent Results Cancer Res 2005;166:277-97.

[38] Baldwin RL, Green JW, Shaw JL, et al. Physician risk attitudes and hospitalization of infants with bronchiolitis. Acad Emerg Med 2005;12:142-6.

[39] Plummer M, Herrero R, Franceschi S, et al. Smoking and cervical cancer: pooled analysis of the IARC multi-centric case--control study. Cancer Causes Control 2003;14:805-14.

[40] Roteli-Martins CM, Panetta K, Alves VA, Siqueira SA, Syrjanen KJ, Derchain SF. Cigarette smoking and high-risk HPV DNA as predisposing factors for high-grade cervical intraepithelial neoplasia (CIN) in young Brazilian women. Acta Obstet Gynecol Scand 1998;77:678-82.

[41] Rajeevan MS, Swan DC, Nisenbaum R, et al. Epidemiologic and viral factors associated with cervical neoplasia in HPV-16-positive women. Int J Cancer 2005;115:114-20.

[42] Kjellberg L, Hallmans G, Ahren AM, et al. Smoking, diet, pregnancy and oral contraceptive use as risk factors for cervical intra-epithelial neoplasia in relation to human papillomavirus infection. Br J Cancer 2000;82:1332-8.

[43] Sasson IM, Haley NJ, Hoffmann D, Wynder EL, Hellberg D, Nilsson S. Cigarette smoking and neoplasia of the uterine cervix: smoke constituents in cervical mucus. N Engl J Med 1985;312:315-6.

[44] Hellberg D, Nilsson S, Haley NJ, Hoffman D, Wynder E. Smoking and cervical intraepithelial neoplasia: nicotine and cotinine in serum and cervical mucus in smokers and nonsmokers. Am J Obstet Gynecol 1988;158:910-3.

[45] Hellberg D, Nilsson S. Smoking and cancer of the ovary. N Engl J Med 1988;318:782-3.

[46] Holly EA, Petrakis NL, Friend NF, Sarles DL, Lee RE, Flander LB. Mutagenic mucus in the cervix of smokers. J Natl Cancer Inst 1986;76:983-6.

[47] Prokopczyk B, Cox JE, Hoffmann D, Waggoner SE. Identification of tobacco-specific carcinogen in the cervical mucus of smokers and nonsmokers. J Natl Cancer Inst 1997;89:868-73.

[48] Waggoner SE, Wang X. Effect of nicotine on proliferation of normal, malignant, and human papillomavirus-transformed human cervical cells. Gynecol Oncol 1994;55:91-5.

[49] Wentz WB, Reagan JW, Fu YS, Heggie AD, Anthony DD. Experimental studies of carcinogenesis of the uterine cervix in mice. Gynecol Oncol 1981;12:S90-7.

[50] Sizemore N, Mukhtar H, Couch LH, Howard PC, Rorke EA. Differential response of normal and HPV immortalized ectocervical epithelial cells to B[a]P. Carcinogenesis 1995;16:2413-8.

[51] Melikian AA, Wang X, Waggoner S, Hoffmann D, El-Bayoumy K. Comparative response of normal and of human papillomavirus-16 immortalized human epithelial cervical cells to benzo[a]pyrene. Oncol Rep 1999;6:1371-6.

[52] Yang X, Jin G, Nakao Y, Rahimtula M, Pater MM, Pater A. Malignant transformation of HPV 16-immortalized human endocervical cells by cigarette smoke condensate and characterization of multistage carcinogenesis. Int J Cancer 1996;65:338-44.

[53] Nakao Y, Yang X, Yokoyama M, Pater MM, Pater A. Malignant transformation of human ectocervical cells immortalized by HPV 18: in vitro model of carcinogenesis by cigarette smoke. Carcinogenesis 1996;17:577-83.

[54] Samir R, Asplund A, Tot T, Pekar G, Hellberg D. Tissue tumor marker expression in smokers, including serum cotinine concentrations, in women with cervical intraepithelial neoplasia or normal squamous cervical epithelium. Am J Obstet Gynecol 2010;202:579 e1-7.

[55] Harris TG, Kulasingam SL, Kiviat NB, et al. Cigarette smoking, oncogenic human papillomavirus, Ki-67 antigen, and cervical intraepithelial neoplasia. Am J Epidemiol 2004;159:834-42.

[56] Lindstrom AK, Ekman K, Stendahl U, et al. LRIG1 and squamous epithelial uterine cervical cancer: correlation to prognosis, other tumor markers, sex steroid hormones, and smoking. Int J Gynecol Cancer 2008;18:312-7.

[57] Lindstrom AK, Stendahl U, Tot T, Hellberg D. Associations between ten biological tumor markers in squamous cell cervical cancer and serum estradiol, serum progesterone and smoking. Anticancer Res 2007;27:1401-6.

[58] Castellsague X, Munoz N. Chapter 3: Cofactors in human papillomavirus carcinogenesis--role of parity, oral contraceptives, and tobacco smoking. J Natl Cancer Inst Monogr 2003:20-8.

[59] Hellberg D, Valentin J, Nilsson S. Long-term use of oral contraceptives and cervical neoplasia: an association confounded by other risk factors? Contraception 1985;32:337-46.

[60] Sikstrom B, Hellberg D, Nilsson S, Brihmer C, Mardh PA. Contraceptive use and reproductive history in women with cervical human papillomavirus infection. Adv Contracept 1995;11:273-84.

[61] Silins I, Kallings I, Dillner J. Correlates of the spread of human papillomavirus infection. Cancer Epidemiol Biomarkers Prev 2000;9:953-9.

[62] Veress G, Csiky-Meszaros T, Czegledy J, Gergely L. Oral contraceptive use and human papillomavirus infection in women without abnormal cytological results. Med Microbiol Immunol 1992;181:181-9.

[63] Chan WK, Klock G, Bernard HU. Progesterone and glucocorticoid response elements occur in the long control regions of several human papillomaviruses involved in anogenital neoplasia. J Virol 1989;63:3261-9.

[64] Yuan F, Auborn K, James C. Altered growth and viral gene expression in human papillomavirus type 16-containing cancer cell lines treated with progesterone. Cancer Invest 1999;17:19-29.

[65] Mitrani-Rosenbaum S, Tsvieli R, Tur-Kaspa R. Oestrogen stimulates differential transcription of human papillomavirus type 16 in SiHa cervical carcinoma cells. J Gen Virol 1989;70 (Pt 8):2227-32.

[66] Correa I, Cerbon MA, Salazar AM, Solano JD, Garcia-Carranca A, Quintero A. Differential p53 protein expression level in human cancer-derived cell lines after estradiol treatment. Arch Med Res 2002;33:455-9.

[67] Lindstrom A, Backstrom T, Hellberg D, Tribukait B, Strang P, Stendahl U. Correlations between serum progesterone and smoking, and the growth fraction of cervical squamous cell carcinoma. Anticancer Res 2000;20:3637-40.

[68] Hellberg D, Lindstrom AK, Stendahl U. Correlation between serum estradiol/progesterone ratio and survival length in invasive squamous cell cervical cancer. Anticancer Res 2005;25:611-6.

[69] Shields TS, Falk RT, Herrero R, et al. A case-control study of endogenous hormones and cervical cancer. Br J Cancer 2004;90:146-52.
      Samir R, Tot T, Asplund A, Pekar G, Hellberg D. Increased serum progesterone and estradiol correlate to increased COX-2 tissue expression in cervical intraepithelial neoplasia. Anticancer Res 2010;30:1217-22.

[70] Samir R, Tot T, Asplund A, Pekar G, Hellberg D. Increased serum progesterone and estradiol correlate to increased COX-2 tissue expression in cervical intraepithelial neoplasia. Anticancer Res 2010;30:1217-22.

2

Cervical Glandular Intraepithelial Neoplasia (CGIN)

Narges Izadi-Mood, Soheila Sarmadi and Kambiz Sotoudeh
Department of Pathology
Tehran University of Medical Sciences,
Iran

1. Introduction

Invasive adenocarcinoma is the second most common malignancy of cervix (after squamous cell carcinoma) and accounts for about 15–25% of all cervical cancers (Hopkins & Morley, 1991). The pre-invasive lesion of the adenocarcinoma of cervix which diagnosed as a spectrum of changes has been named cervical glandular intraepithelial neoplasia (CGIN). Over the past several decades, the incidence of cervical adenocarcinoma as well as its relative proportion to squamous cell carcinoma has been increasing. In 1950s and 1960s, adenocarcinoma accounted for only 5% of all invasive cancers of cervix, however it was increased and responsible for 10-22% in 1990s (Hopkins & Morley,1991; McCluggage,2000; Zaino, 2000, 2002; Wang et al.,2004; Leminen et al.,1990).

This increment may be representing both a real and an apparent increase due to a reduction in the number of invasive cervical squamous carcinomas as a consequence of organized screening programs. There also may be due to better recognition of adenocarcinoma and dysplastic endocervical glandular lesions by pathologists and appreciation of the fact that some poorly differentiated carcinomas may be glandular rather than squamous in type (revealed by the use of ancillary staining), therefore favors a real increase in the incidence of adenocarcinoma in women below 35–40 years of age (McCluggage, 2000).

The pre-invasive lesions of cervical adenocarcinoma are a heterogeneous group with various histomorphological patterns which may be confused with a wide range of non-neoplastic glandular lesions; therefore it is imperative to recognizing these presumed precursors as well as knowledge of their differential diagnosis.

This chapter focuses on an overview of the different terminology, various histopathological features, ancillary diagnostic techniques, and practical diagnostic approach to cervical glandular intraepithelial neoplasia.

2. Precursors glandular lesion of the uterine cervix

2.1 Definition

Cervical glandular intraepithelial neoplasia (CGIN) is a spectrum of presumed pre-invasive (or preneoplastic) cervical glandular lesion. The term 'presumed pre-invasive' is used because there is some controversy as to whether these lesions, especially at the lower end of the spectrum, progress to adenocarcinoma (McCluggage, 2000). The concept of histological recognizable pre-invasive form of adenocarcinoma was at first suggested by Friedell and McKay in 1953. They have proposed that like other organs such as breast, stomach, bronchus, skin and also squamous cell carcinoma of cervix, adenocarcinoma of cervix could have these precursor lesions. Subsequent investigation was renewed interest in characterizing precursor lesions of invasive adenocarcinoma with intent to invoke a unifying theory of a common subcolumnar reserve cell for all types of cervical cancer or to categorize lesions in a fashion analogous to precursors of squamous cell carcinoma of the cervix (Zaino, 2002; Christopherson, 1979). Smedts et al. had reported that cervical intraepithelial neoplasia (CIN), combined adenocarcinoma in situ (AIS) / CIN, and a part of the solitary AIS lesions share a common, marker phenotype comparable with that endocervical reserve cells, which is indicate of a common origin. However, a second group of solitary AIS lesions with an endocervical phenotype possibly originate from a luminal type progenitor cells, within the endocervix (Smedts et al., 2010). Although endocervical adenocarcinoma in situ (AIS) is a known precursor of invasive adenocarcinoma, there is no universally accepted precursor lesion of AIS itself (Ioffe et al., 2003).

2.2 Classification and terminology of pre-invasive cervical glandular lesions

There is no consensus about the terminology using for the classification of pre-invasive endocervical glandular lesions. The term of adenocarcinoma in situ (AIS) as a precursor lesion of adenocarcinoma of uterine cervix was first described by Friedell and McKay in 1953 and most subsequent studies used this terminology (Friedell & Mckay, 1953). Other terms used to describe pre-invasive endocervical glandular lesions include *endocervical glandular dysplasia, cervical intraepithelial glandular neoplasia, cervical glandular atypia, endocervical columnar cell intraepithelial neoplasia* and *atypical endocervical hyperplasia*.

The International Society of Gynecological Pathologists under the auspices of the World Health Organization (WHO) included categories of glandular atypia, atypical hyperplasia (glandular dysplasia), adenocarcinoma in situ and invasive adenocarcinoma in its classification (McCluggage, 2000; Kurman, 2010).

In WHO classification, 3 categories were introduced: *1-Glandular atypia:* which refers to nonneoplastic changes often associated with inflammation; *2-Atypical hyperplasia (glandular dysplasia):* which refers to intraepithelial glandular neoplasia that is less severe than AIS and *3- AIS* (Kurman,2010).

Cervical glandular intraepithelial neoplasia (CGIN) : which is a three-tier grading system *(CGIN 1, 2 and 3)*similar to that used for pre-invasive cervical squamous lesions, that originally has been introduced by Gloor and Hurlimann (Gloor & Hurlimann,1986).This three-tier grading system was performed according to cytohistological criteria including nuclear abnormality, presence of mitosis, amount of intracellular mucin and architectural abnormality. Following this grading a new terminology was introduced by a working party of the Royal College of Pathologists and the NHS Cervical Screening Program in the Britain (NHS Cervical Screening Programme [NHSCSP], 1999). Because of difficulties in three-tier grading, particularly the distinction between CGIN 1 and 2, most authors therefore recognize only two grades of CGIN, termed high grade and low grade CGIN. This does not mean that

CGIN is a two stage disease but reflects the fact that differentiation into three grades is probably poorly reproducible. Alternatively high grade and low grade CGIN may be designated as AIS and glandular dysplasia, respectively (McCluggage, 2000).

The Silverberg group (Ioffe et al.,2003) introduced the Silverberg scoring system for assessment of the endocervical glandular lesions that is designed to aid in diagnosis and to bring about better inter- and intra observer agreement in this difficult area (McCluggage, 2000; Liang et al.,2007). This scoring scheme is based on 3 separately graded components: nuclear stratification, nuclear atypia, and mitoses/apoptosis. The scores for which are added to result in the total score equivalent to a diagnostic category: benign (score = 0-3), endocervical glandular dysplasia (score = 4-5), and adenocarcinoma in situ (score = 6-9) (Table 1) (Ioffe et al., 2003).

Stratification
-None = 0
-Mild = 1
-Moderate = 2
-Up to the luminal surface = 3
Nuclear atypia
-As normal = 0
- Small (size of normal) or slightly enlarged uniform nuclei, minimal hyperchromasia, little dispolarity, no nucleoli = 1
-Nuclear enlargement (up to 3 × normal), moderate anisocytosis, moderate hyperchromasia, moderate dispolarity, occasional small nucleoli = 2
-Large nuclei (>3 × normal), marked anisocytosis, marked hyperchromasia, severe dispolarity, frequent prominent nucleoli = 3
Mitoses and apoptosis
[In two most active glands, number per gland (average between two glands)]
-None= 0
-Less than 0.5 per gland = 1
-0.6–3.0 per gland = 2
- >3.0 per gland = 3
Total score
0–3 = benign
4–5 = endocervical glandular dysplasia (EGD)
6–9 = adenocarcinoma in situ (AIS)

Table 1. Silverberg group's scoring system for assessment of the endocervical glandular lesions

2.3 Pathogenesis

Among a variety of factors investigated, including the absence of a prior Pap smear, number of sexual partners, age at first intercourse, history of genital infections, obesity, and tobacco use, two conditions have emerged as potential risk factors in the development of cervical adenocarcinoma: Human Papilloma Virus (HPV) infection and oral contraceptive (OCP) use (Zaino,2000). But from different descriptive epidemiological observations, it has been suggested that adenocarcinoma may differ in pathogenetic mechanisms and that its etiology should be investigated with reference to hormonal, rather than infectious, aspects (Parazzini et al., 1988). Ursin et al. reported that the highest risk was for oral contraceptive use for more than 12 years. No additional increased risk was found for early age at start of oral contraceptive use, use before age 20 or before first pregnancy, time since first use, time since last use, or particular formulations, once total duration of use had been accounted for (Ursin et al., 1994).But in the study by Parazzini et al., oral contraceptive use was not related to the risk of adenocarcinoma of the cervix (Parazzini et al., 1988; Madeleine et al., 2001).

Although morphologic evidence of productive HPV infection is generally limited to squamous or transitional epithelium, now overwhelming data supports the high frequency of HPV infection in both AIS and invasive adenocarcinoma (Madeleine et al., 2001; Bulk et al., 2006). In early 1980s wart viruses were not found in many of the in situ and invasive adenocarcinoma of the cervix, but with more sensitive techniques, HPV type 16, 31, and more frequently 18, have been identified in 80% and more of adenocarcinoma and adenosquamous carcinoma (Zaino, 2000). Recent studies have shown that HPV type 18 and 16 are the most common types which are detected in 43% and 23% of CGIN, respectively (Zielinski et al.,2003; Pirog et al.,2000).

2.4 Clinical signs and colposcopic features

The early diagnosis of glandular lesions still represents a real challenge for clinicians, who are likely to miss the lesions because of the absence of clinical indicators, normal cytology, or cytology suggestive of squamous disease and/or because of unfamiliarity with the diseases newly delineated colposcopic presentations.

For identifying pre-invasive cervical glandular lesions, colposcopy has not been helpful since colposcopic features of AIS and early adenocarcinoma are widely seen as known nonspecific and also this is because the disease only slightly changes the surface contour and because the neoplastic "glands "are often buried beneath the surface (Campion, 2010; Wright, 2002).

Usually most glandular lesions lie within or close to the transformation zone. While the majority of squamous lesions are usually visible by colposcopic examination, AIS may locate proximally, involving the endocervix, or may lie under the metaplastic epithelium or placed in an abnormal transformation zone and thus be out of colposcopic view (Campion, 2010; Wright, 2002).

Because of AIS coexists with high grade CIN in 30%-70% of cases and the location of the lesion, the abnormal smear will frequently predicate only the squamous lesion (van Aspert-van Erp et al.,2004). In mixed conditions that AIS and squamous cell lesions are concomitantly present, cytologic examination may only exhibiting squamous abnormality that mislead the colposcopist to look exclusively for a squamous lesion and to be satisfied upon finding it. Furthermore, the colposcopic biopsy may confirm the squamous lesion, with AIS being detected only on a subsequent wide excision or within a hysterectomy specimen. Diagnostic excisional biopsy must be always performed when AIS is found on punch biopsy or when AIS is suspected cytologically or colposcopically but not proven histologically (Campion, 2010).

Commonly colposcopic diagnosis of glandular lesions is less than satisfactory because have no specific appearance and mostly mimic the other lesions.

However, to overcome this problem a new set of colposcopic criteria has been recommended for differentiation between glandular lesion and metaplasia, condyloma, squamous intraepithelial neoplasia and squamous cell carcinoma.

The criteria are:

- Lesion locatied over columnar epithelium, not contiguous with the squamocolumnar junction;
- Large "gland" or crypt openings;
- Papillary structure;
- Budding;
- Patchy red and white coloration;
- Waste-thread, tendril, root, and character-writing blood vessels;
- Single or multiple dots produced at tips of papillary projection by looped vessels (Wright, 2002; Ostör et al., 1984).

Some features can be used to eliminate a lesion from consideration, such as punctuation and true mosaic pattern (which are present only in squamous intraepithelial lesions) and corkscrew vessels (which are associated only with invasive squamous disease).

Although many colposcopically recognized features are common to a variety of diseases, paying attention to surface contour and vascular configurations can greatly help the colposcopist discover glandular disease when it is present and differentiate it from other conditions (Wright, 2002).

2.5 Morphological features

2.5.1 Histopathological features of cervical glandular intraepithelial neoplasia (CGIN)

Adenocarcinoma in situ (AIS) of the cervix has no distinguishing clinical and colposcopic features, and because it is rare, pathologists may not be familiar with its microscopic appearances. It is easily overlooked since it may be focal and because it is frequently associated with cervical intraepithelial neoplasia (CIN), which is more impressive. There is little information about its natural history (Ostör et al., 1984).

Although there are several proposed classifications and terminologies used for describing CGINs, in most of them the morphologic criteria are nearly the same with small differences, set them in a wide spectrum between the reactive/benign lesions and invasive carcinoma in two extremities of classification.

By using these diagnostic criteria, identification of high grade lesion has been more reproducible than low grade , but most often the diagnosis of low grade lesion resulted in a confusing state of affairs for pathologists and clinicians. In most instances high grade CGIN (HCGIN) is diagnosed more often than low grade CGIN (LCGIN) in contrast to some earlier studies that reported low grade CGIN was more common (Brown, 1986).

There is a popular misconception among pathologists and gynecologists that CGIN often occurs in upper parts of the endocervical canal. However, in most, but not all cases, CGIN occurs close to the transformation zone (McCluggage, 2003). CGIN is commonly associated with a concomitant squamous intraepithelial lesion and may affect the surface epithelium and/or endocervical crypts, usually in the region of the transformation zone (Bekkers et al., 2003).

Another popular misconception is that skip lesions are extremely common in CGIN. Skip lesions undoubtedly do occur but these are relatively uncommon, probably occurring in up to 15% of patients (McCluggage, 2003).

Both high and low grades CGIN are characterized by a combination of cytological and architectural features. These features are more pronounced in high grade CGIN but not all are necessarily present in a given case of CGIN (McCluggage, 2000).

The recognition of low grade CGIN is more problematic and this lesion easily underdiagnosed by pathologists. Low grade CGIN has many overlapping features with reactive changes. Dysplastic changes in low grade CGIN are not fulfill the diagnosis of high grade CGIN and qualitatively less severe than high grade CGIN. The most common changes are: glands composed of pseudostratified cells that slightly loss their polarity, with large and hyperchromatic nuclei and minimal mitosis and apoptotic bodies. Usually stroma lacks any inflammatory or reactive changes (NHSCSP, 1999) (Figure 1- A and B).

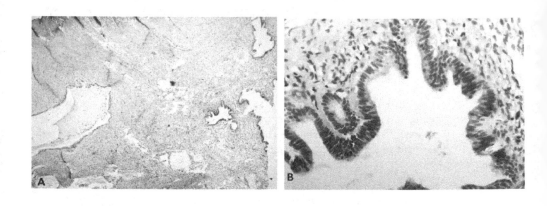

Fig. 1. LCGIN. **A,** Normal crypt in the left comparable with darker epithelium of LCGIN in the right in low magnification. **B,** Partially pseudostratified epithelium, crowded nuclei, and occasional mitotic figures are seen in higher magnification.

In high grade CGIN which abruptly begins beside the normal columnar cells (Figure 2- A and B) dysplastic changes are more severe and characterized by usually crowded glands (Figure 2- C) with various architectural patterns like budding and branching (Figure 2- D), exophytic papillae (Figure 3-A), intraluminal papillary projections (Figure 3-B) , micropapillae (Figure 3-C) and cribriform (Figure 4-A and B) , that composed of atypical cells that variably loss their cytoplasmic mucin, and display large pleomorphic hyperchromatic nuclei with lack of polarity and easily finding mitoses and apoptotic bodies (Figure 4- C and D)(For detailed definition of histopathological features see Table 2).Glands are commonly surrounded by a compact stroma (NHSCSP, 1999).

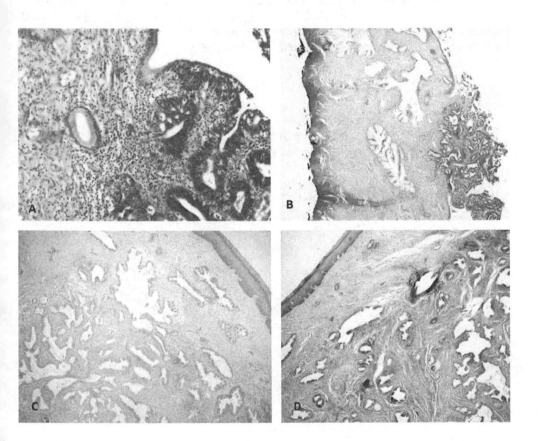

Fig. 2. HCGIN. **A,** Abrupt transition from normal endocervical columnar epithelium (top & left) to the stratified epithelium of the CGIN (right). **B,** Abrupt transition from normal endocervical columnar epithelium (top) to the stratified epithelium of the CGIN (center). **C,** A cluster of closely packed glands with branching, out-pouching and occasional in-folding. **D,** A cluster of glands with branching (irregular contour).

High grade CGIN displays three common histological subtypes: endocervical, endometrioid and intestinal as well as several uncommon subtypes: serous, clear cell, adenosquamous, villoglandular and tubal (McCluggage, 2000; Zaino, 2000).

Among them the endocervical HCGIN (alone or admixed with other types) is the most common type which mimic the normal endocervical glands. In contrast to endocervical type that small to moderate amount of cytoplasmic mucin present in luminal side of atypical cell, in endometrioid HCGIN which cells mimic the proliferative endometrial cells, their eosinophilic cytoplasm lack any mucin. Another characteristic feature in this type is significant nuclear pseudostratification. Intestinal HCGIN is recognized by its prominent goblet cell and occasional neuroendocrine or paneth cells. There is no evidence that behavior of the different subtypes of HCGIN is significantly differed (McCluggage, 2000; Zaino,2000,2002; Wang et al., 2004; Gloor & Hurlimann,1986; NHSCSP, 1999; Brown & Wells,1986; Gloor & Ruzicka,1982; Kurian & al-Nafussi,1999).

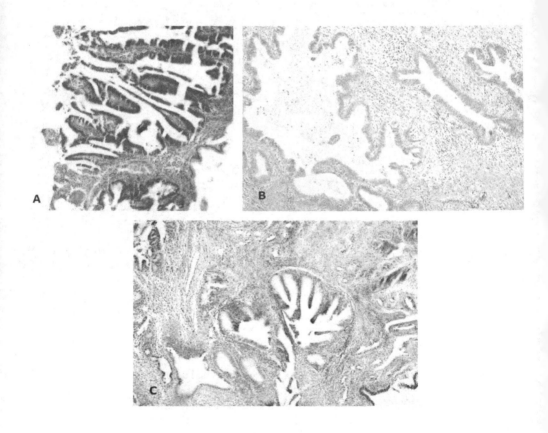

Fig. 3. HCGIN. **A,** Simple exophytic papillary pattern with thin delicate stromal stalk.
B, Infolding of epithelium into the glandular lumina with supporting stroma.
C, Intraluminal exuberance and delicate micropapillary projection with no supporting stroma.

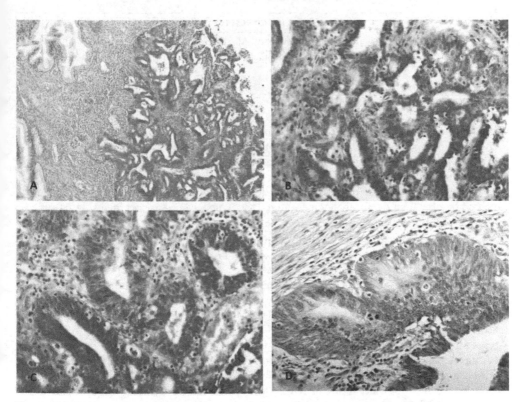

Fig. 4. HCGIN. **A,** Macroglands with secondary or multiple generation of bridging subdividing the lumen into smaller glandular spaces; no stroma supports the bridging cells (low power). **B,** Macroglands with secondary or multiple generation of bridging subdividing the lumen into smaller glandular spaces; no stroma support the bridging cells (High power). **C,** Nuclear stratification (loss of mucin secretion); nuclear hyperchromasia with mitotic activity and apoptotic bodies. **D,** Nuclear Stratification; loss of nuclear polarity with mitotic activity and apoptotic bodies.

2.5.2 Cervical cytology

The diagnostic category and the terminology of atypical glandular cell (AGC), has been widely used since it was first established at the 2001 Bethesda convention (Covell et al., 2003). Before 2001, AGC within The Bethesda System (TBS) were mentioned as atypical glandular cells of undetermined significance (AGUS). The incidence of endocervical adenocarcinoma has increased steadily over the past two decades (Hopkins & Morley, 1991). Since TBS was introduced, the diagnosis of AGC has risen and now accounts for 0.17–1.83% of all cervical smears (Nasu et al., 1993). The term AGC applies to glandular cells that demonstrate changes beyond those typical of benign reactive processes but lack sufficient features for a diagnosis of adenocarcinoma. Generally, the origin of AGCs, endocervical or endometrial, can be distinguished based on the larger nuclear size and more abundant cytoplasm of endocervical cells (Solomon et al., 1998).

Definition of architectural features
-Glandular crowding: A cluster of closely packed glands.   **-Glandular budding:** Glands out-pouching into the surrounding stroma to produce "finger-in-glove" pattern.   **-Glandular branching:** Glands with multiple out-pouching and irregular counters.   **-Villoglandular /exophytic papillae:** Simple branching exophytic papillary pattern with thin delicate stromal stalk,reminds of villous adenoma of the GI tract.   **-Intraluminal papillary projections:** Infolding of epithelium into the glandular lumina with supporting stroma which creates a cribriform-like pattern.   **-Micropapillae pattern:** Intraluminal exuberance and delicate micropapillary projection with no supporting stroma.   **-Cribriform pattern:** Macroglands with secondary or multiple generation of bridging subdividing the lumen into smaller glandular spaces; no stroma support the bridging cells.
Definition of cytological features
-Abrupt junction between the normal columnar epithelium and the CGIN: Partially affected epithelial lining by CGIN with sharp demarcation between normal epithelium and CGIN.   **-Intestinal metaplasia /goblet cell formation:** A form of intestinal metaplasia, exhibits goblet cells and even paneth or neuroendocrine cells.   **-Loss of mucin secretion in cells of endocervical type:** Reduction or complete absence of intracellular mucin.   **-Nuclear stratification:** Pseudostratified up to stratified epithelium in reciprocal reduction in cytoplasmic mucin with or without loss of nuclear palisading and polarity.   **-Nuclear changes:** Enlarged, elongated, pleomorphic and hyperchromatic nuclei with granular dense and evenly or abnormal dispersion of nuclear chromatin and presence of prominent nuclei.   **-Mitotic activity:** Presence of juxtaluminal and increased numbers of mitotic figures (normal/ or abnormal).   **-Apoptosis (apoptotic bodies):** Markedly condensed homogenous nuclei (with or without nuclear fragmentation) often associated with densly eosinophilic cytoplasm.

Table 2. Definition of various architectural and cytological features of CGIN

The Bethesda System (TBS) recommends subclassification of AGC into the categories of *"favor endocervical origin"*, *"favor endometrial origin"* and *"not otherwise specified (NOS)"*. Favor endocervical origin lesions are further classified into the categories of *"favor neoplastic"* and *"favor NOS"* (Covell et al., 2003). However, subclassification of AGC has yet to be proved clinically effective, and although The Bethesda Committee and many others have studied cytologic criteria important in subclassification, these criteria have not been tested vigorously. The rates of AGC (reported as AGUS before 2001) quoted in the literature vary from 0.095% to 1.83% (Nasu et al., 1993; Mood et al. , 2006; Marques et al.,2011; Tam et al.,2003; Scheiden et al.,2004; Pecorelli et al., 2009).

According to TBS 2001, cytological features of subcategorized AGC is as follow:

In atypical endocervical cells (NOS) (Figure 5-A and B):

- Cells occur in sheets and strips with some crowding and nuclear overlap.

- Nuclear enlargement, up to three to five times the area of normal endocervical nuclei, may be seen.
- Some variation in nuclear size and shape is present.
- Mild hyperchromasia is frequently evident.
- Nucleoli may be present.
- Mitotic figures are rare. Cytoplasm may be fairly abundant, but the nuclear /cytoplasmic (N/C) ratio is increased.
- Distinct cell borders often are discernible.

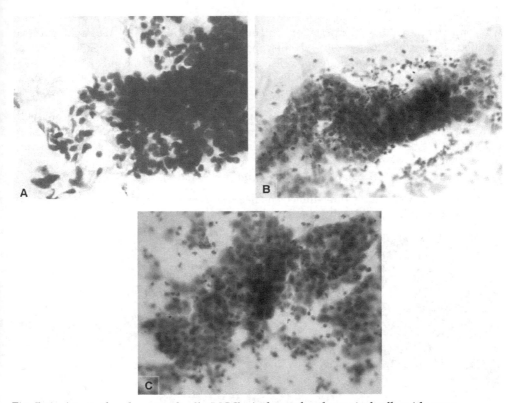

Fig. 5. **A**, Atypical endocervical cells (NOS) .A sheet of endocervical cells with some crowding and nuclear overlap. **B**, Atypical endocervical cells (NOS). Strip of endocrvical cells with stratification, elongation of nuclei, nuclear enlargment and hyperchromasia . **C**, Atypical endocervical cells, (favor neoplastic). A sheet of endocervical cells with crowding and nuclear overlap shows increased nuclear/cytoplasmic ratios.The quantity of cytoplasm is diminished, and cell borders are ill defined.

In liquid-based preparation groups are more rounded and three dimensional with piled-up of cells, making individual cells in the center difficult to visualize.

In atypical endocervical cells, (favor neoplastic) (Figure 5-C) alongside above mentioned features, added cytological features incorporated:

- Cell morphology, either quantitatively or qualitatively, falls just short of an interpretation of endocervical adenicarcinoma in situ or invasive adenocarcinoma.
- Rare cell groups may show rosetting or feathering
- Nuclei are enlarged with some hyperchromasia
- Occasional mitosis may be seen.
- Nuclear/cytoplasmic ratios are increased, quantity of cytoplasm is diminished, and cell borders may be ill defined.

In liquid-based preparation groups may be three dimensional, thick, with layers of cells obscuring central nuclear detail.

It is important that the interpretation of "atypical glandular cells" (AGC) should be qualified, if possible, to indicate whether the cells are thought to be endocervical or endometrial origin. If the origin of the cells cannot be determined, the generic "glandular" term is used. Atypical endocervical cells should be further qualified when a particular entity, including neoplasia, is favored.

AIS is often identified in cytological specimens as abundant abnormal cells, typically with columnar configuration, single cells, two-dimensional sheets, or three-dimensional clusters and syncytial aggregates with nuclear crowding and overlap, without an accompanying tumor diathesis. Characteristic features of glandular differentiation include rosette formation, nuclear feathering ,and palisading (Figure 6-A and B).

In liquid-based preparation three-dimensional clusters are common. Chromatin is more open (vesicular) with irregular distribution and parachromatin clearing (Covell et al., 2003).

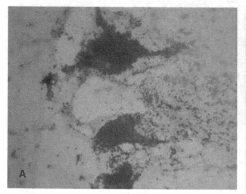 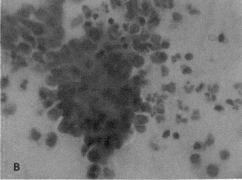

Fig. 6. Adenocarcinoma in situ. **A,** Abundant abnormal cells, typically with columnar configuration, single cells, two-dimensional sheets, or three-dimensional clusters and syncytial aggregates with nuclear crowding and overlap. **B,** Characteristic rosette formation in glandular differentiation.

2.5.3 Differential diagnosis

2.5.3.1 Invasive adenocarcinoma

The most important differential diagnosis of CGIN is microinvasive and invasive adenocarcinoma.

By definition the concept of microinvasive adenocarcinoma (MIA) is the same as microinvasive squamous cell carcinoma and represents an invasive adenocarcinoma with limited depth of stromal invasion up to 5 millimeters (Pecorelli et al., 2009). Despite plenty data about its squamous counterpart, MIA is suffering from a reliable cytological and histological diagnostic criteria as well as information about its prognosis and management. While diagnosis of early stromal invasion in squamous cell carcinoma (SCC) is relatively simple and easy, however, identifying early invasion in high grade CGIN may be extremely difficult or even impossible (McCluggage, 2000; Zaino, 2000; NHSCSP, 1999,Nucci,2002). High grade CGIN should be limited to the normal glandular field but problems occur when closely packed, architecturally abnormal glands are lined by dysplastic epithelium which fulfils the criteria for a diagnosis of high grade CGIN (McCluggage, 2000; Zaino, 2000; Nucci, 2002).

MIA characterized by effacement of normal glandular tissue by irregular atypical glands that extends beyond the deepest normal crypt associated with a stromal reactivity of desmoplastic, infiltration of chronic inflammatory cells or edematous type.

There are two certain features that identify the presence of invasion in endocervical adenocarcinoma (1) Individual cells or incomplete glands (Figure 7-A and B) lined by cytologically malignant-appearing cells at a stromal interface and (2) malignant appearing glands surrounded by a host response (Figure 7-C). It is important to determine that the glands are lined by cytologically malignant-appearing cells, because endocervicitis, microglandular hyperplasia, and ruptured mucin-filled glands all may have incomplete glands that at times may be associated with a host response of dense inflammation and, occasionally, edema or fibrosis. Unfortunately, many adenocarcinomas do not display these two changes yet are invasive. It should be noted that infiltration of chronic inflammatory cells around the CGIN may also be present and result in a confusing and complex status (McCluggage, 2000; Zaino, 2000; Nucci, 2002).

Additional features that are not entirely specific may help to identify invasion in other cases including:

(1) Architecturally complex, branching, or small glands, which grow confluently or in a labyrinthine pattern; (2) A cribriform growth pattern of malignant-appearing epithelium devoid of stroma within a single gland profile; (3) The presence of glands below the deep margin of normal glands; and (4) The presence of early stromal infiltration from glands involved by HCGIN of small buds of cells, often with a squamoid appearance (McCluggage, 2000; Zaino, 2000; Nucci, 2002).

Large masses of densely packed architecturally complex glands with luminal bridges and a cribriform growth pattern strongly suggest invasion. More difficult is the assessment of the "deep margin" of normal glands. Although it is stated that endocervical glands should be confined to the inner third of the cervix and less than 1 cm deep, benign glands in various patterns including nabothian cysts, tunnel clusters, laminar endocervical hyperplasia, deep endocervical glands, and mesonephric duct remnants may be found deeper in the stroma on occasion. Pathologists should, wherever possible, make every effort to make this distinction, but it is recognized that there will be cases in which the pathologist remains uncertain as to whether a lesion is invasive or not, even after the mandatory examination of many levels. This must be stated in the report (McCluggage, 2000, 2003; Zaino, 2000; NHSCSP, 1999; Nucci, 2002).

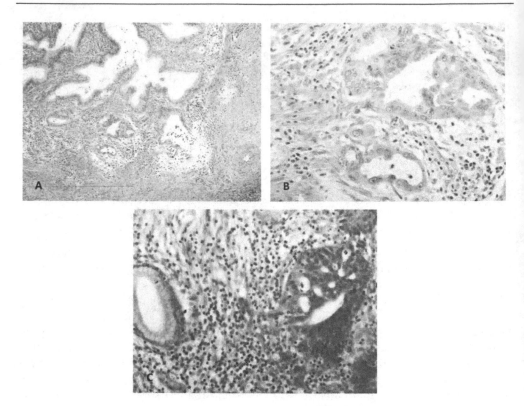

Fig. 7. Microinvasive adenocarcinoma. **A,** Individual cells and incomplete glands beneath the crypt surronded by edematous stroma and lymphocytic infiltration (low power). **B,** Individual cells and incomplete glands beneath the crypt surrounded by edematous stroma and lymphocytic infiltration (High power).Note the severe degree of cytological atypia. The nuclei are pleomorphic, there is loss of nuclear polarity and several nuclei contain large nucleoli. Cell above the incomplete gland has copious eosinophilic cytoplasm. **C,** A cribriform macrogland with cells in the lower right part with copious eosinophilic cytoplasm should arouse a suspicion that invasion may be present. Note normal gland in the left.

2.5.3.2 Tuboendometrioid metaplasia and Endometriosis

Tuboendometrioid metaplasia (TEM) is very common within the cervix and the most common lesion to be misdiagnosed as CGIN (McCluggage, 2000). It usually develops after cervical biopsy or diathermy, but may also occur in the absence of any surgical intervention. In TEM, the normal endocervical surface or crypt epithelium replaced by tubal or endometrioid cell type or by a population of cubo-columnar cells with regular, oval to round, darkly staining, hyperchromatic basal nuclei and high nuclear/cytoplasmic ratios; some of the cells may be ciliated (Figure 8- A and B). Tubal metaplsia usually involves a single gland or a few glands near the squamocolumnar junction and is not associated with inflammation.

Mitoses are uncommon except when estrogenic proliferative activity is present. Nuclear pleomorphism and atypical mitoses are absent (NHSCSP, 1999). A helpful clue to the diagnosis is the presence of cilia on the luminal border of some of the cells.

Occasionally invasive cervical adenocarcinoma may also contain ciliated cells, thus it is not need to overemphasize that the presence of cilia in the cervix does not unequivocally denote a benign process.

Ancillary techniques, such as the use of proliferation markers, have been used with some success in attempting to distinguish TEM and other benign glandular lesions from CGIN. These are discussed in detail later (McCluggage, 2000).

Endometriosis, which is characterized by the presence of endometrial-type glands set in an endometrial stroma, most commonly occurs in the region of the external cervical os or in the lower endocervical canal (Figure 9-A and B).

At colposcopy it appears as a hemorrhagic lesion. Regular bleeding may lead to stromal fibrosis and stenosis of the external cervical os. It can usually be easily recognized histologically and, if active, is most commonly approximately in phase with the intrauterine endometrium (NHSCSP, 1999) (Figure 9-C).

Cervical endometriosis not associated with TEM may have either a superficial or deep location. Deep cervical endometriosis is often associated with pelvic endometriosis and generally causes no problems in diagnosis. However, superficial endometriosis may be mistaken for CGIN (Tam et al., 2003). The presence of endometrial type stroma is a clue to the diagnosis but this is often significantly obscured by accompanying inflammation or hemorrhage and rarely by smooth muscle metaplasia. Particularly in young women there may be considerable mitotic activity when estrogen induced proliferative activity is present (McCluggage, 2000) (Figure 9-D).

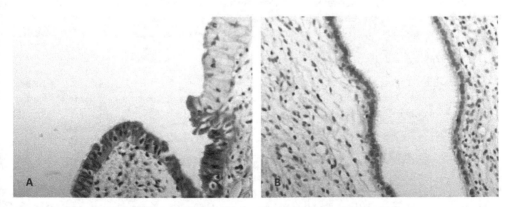

Fig. 8. **A,** Tubal metaplasia. Note abrupt transition between the mucus-secreting endocervical cells and the ciliated cells. **B,** Tuboendometrioid metaplasia (TEM). The normal endocervical crypt epithelium replaced by a population of cubo-columnar cells with regular, oval to round, darkly staining, hyperchromatic basal nuclei and high nuclear/cytoplasmic ratios; some of the cells may be ciliated.

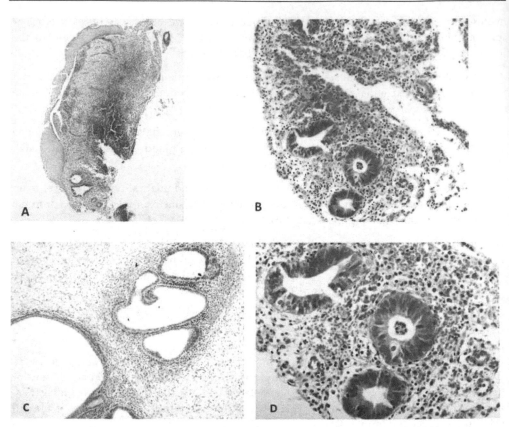

Fig. 9. Endometriosis. **A,** Presence of endometrial-type glands set in an endometrial stroma beneath the squamous epithelium (low power). **B,** High power view. **C,** Subnuclear vacuoles in active endometriosis. **D,** Note mitotic activity in estrogen induced proliferative activity.

2.5.3.3 Microglandular hyperplasia

Microglandular hyperplasia or microglandular adenosis is a common lesion, seems to be a result of progesterone effects, and most commonly found in pregnant women or those receiving oral contraceptives or progestines.

In gross findings is often polypoid and may be unifocal or multifocal (Fig. 10- A).

Early lesions may show sessile. Microscopically, it is characterized by the presence of closely packed small glandular structures lined by cuboidal epithelial cells with vesicular nuclei. Mitotic figures are uncommon, but may be found, and there is often prominent subnuclear and supranuclear vacuolation. There may be associated with reserve cell hyperplasia and immature squamous metaplasia and there is often a striking neutrophilic infiltrate (Figure 10- B and C).

Signet ring like cells may be seen. Typical or atypical forms of microglandular hyperplasia may be mistaken for CGIN or clear cell carcinoma. The suspicion of malignancy may be heightened when microglandular hyperplasia results in a polypoid mass. CGIN and clear cell carcinoma generally have a higher mitotic rate than microglandular hyperplasia, atypical mitoses are often seen and nuclei are not vesicular (McCluggage, 2000; Zaino, 2000, 2002; NHSCSP, 1999; Ostör, 2000, Nucci, 2003).

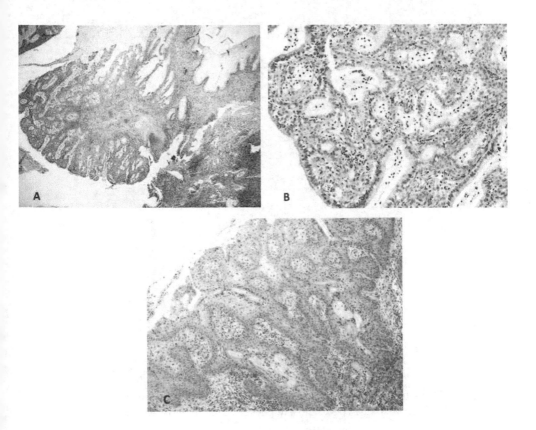

Fig. 10. Microglandular hyperplasia. A, Polypoid configuration (low power). B, Note small size glands and neutrophilic infiltration. C, Reserve cell hyperplasia and immature squamous metaplasia.

2.5.3.4 Tunnel clusters

Tunnel clusters are benign, relatively rare, endocervical lesions which are most common in multigravid patients. This has led some to suggest that they are a result of subinvolution of endocervical glands following pregnancy. Tunnel clusters are characterized by a lobular arrangement of closely packed, often dilated endocervical glands. The lining epithelium is of mucinous type but is often compressed and attenuated and filled by mucinous eosinophilic secretions. Nuclear pleomorphism and mitotic figures are absent and the lesion is always an incidental finding. Two histological types have been described. In type A there is little or no dilatation of glands whereas type B is characterized by marked glandular dilatation (Figure 11).

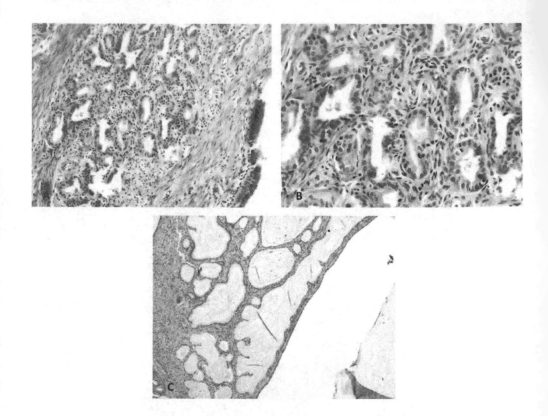

Fig. 11. Tunnel clusters. **A,** Type A. Closely packed endocervical glands (Low power). **B,** Type A. (High power). **C,** Type B. Closely packed, dilated endocervical glands.

Although malignancy, especially minimal deviation adenocarcinoma (or adenoma malignum), may be considered, this is not a significant problem and once the characteristic histological features of tunnel clusters are known, they are easily appreciated (McCluggage, 2000; Zaino, 2000, 2002; NHSCSP, 1999; Ostör, 2000, Nucci, 2003).

2.5.3.5 Reactive glandular atypia

This very common category, including atypia as a result of inflammation, tissue repair, and response to irradiation, may mimic adenocarcinoma. Inflammation and tissue repair may result in lacelike masses of glandular cells with enlarged, pleomorphic nuclei and prominent nucleoli. Typically, the epithelium lining the glands is not stratified. The presence of a dense inflammatory infiltrate, frequently extending into the epithelium, often coupled with loss of polarity and acquisition of abundant, polygonal cytoplasm, assists in the recognition of the presence as reactive. Isolated multinucleated endocervical cells are common (Figure 12).

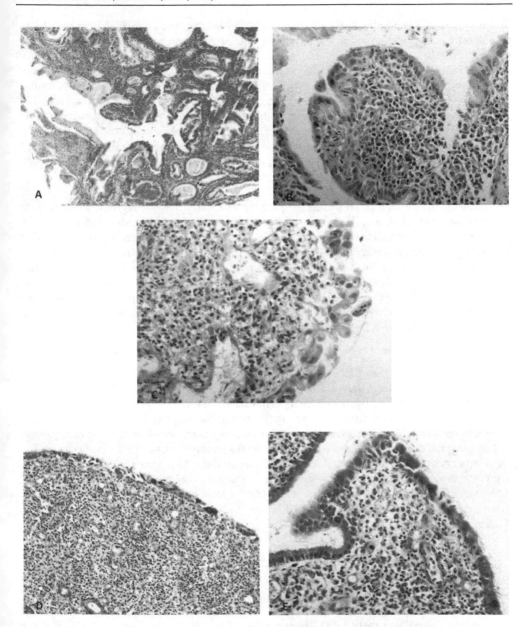

Fig. 12. Reactive glandular atypia. **A,** Note loss of polarity in a part of crypt lining produce abrupt transition and darker area between normal and reactive epithelium. **B,** Note enlarged, pleomorphic nuclei in endocervical epithelium. **C,** Dense inflammatory infiltrate, extending into the epithelium, coupled with loss of polarity.Enlarged, pleomorphic nuclei in endocervical epithelium and dense inflammatory infiltrate in the stroma. **D,** (Low power). **E,** (High power).

Reactive atypia is generally differentiate from CGIN by the lack of epithelial stratification, degenerative or reactive type changes in nuclear chromatin rather than granular hyperchromasia, and a paucity of mitotic activity and apoptotic bodies. Papillary endocervicitis is a specific form of tissue response characterized by relatively short edematous papillae, often containing lymphoid aggregates, covered by a simple columnar epithelium displaying nuclear changes of reactive cells. In contrast, radiation may result in glands being lined by large columnar or cuboidal cells with very large, hyperchromatic nuclei, but the chromatin is usually smudged and mitoses are rare. A clue to the reactive nature is that the abrupt transition to normal endocervix commonly seen in CGIN is not present (McCluggage, 2000; Zaino, 2000, 2002; NHSCSP, 1999).

2.5.3.6 Arias-Stella reaction

The Arias-Stella reaction is an incidental finding in about 10% of pregnant women. This reaction may involve endocervical glands as well as cervical endometriosis during pregnancy. The histological appearances of cells with enlarged pleomorphic nuclei and abundant vacuolated clear or eosinophilic cytoplasm are well known but may be misdiagnosed as CGIN or clear cell carcinoma. The fact that this lesion is focal and associated with the history of pregnancy facilitates the diagnosis. Mitotic figures are uncommon but may occur in the Arias-Stella reaction and indeed the presence of abnormal mitotic figures has been described in the Arias-Stella reaction involving endometrial glands (McCluggage, 2000; Zaino, 2000, 2002; NHSCSP, 1999).

2.5.3.7 Mesonephric remnants and hyperplasia

Mesonephric remnants take place in up to 22% of cervices. Their occurrence varies with the type of specimen because they are seldom seen in biopsy specimens, but are relatively common in conization and hysterectomy specimens in which deep portions of the cervix are routinely examined. The mesonephric or Wolffian ducts commonly persist as small remnants usually located in the lateral walls of the vagina or cervix, in the broad ligament, and in the hilus of the ovary. Microscopic lobules frequently surround a central duct within the deep cervical stroma. The acini are lined by cuboidal cells with oval, bland nuclei and scant to moderate quantities of eosinophilic cytoplasm. Mucin is not present in the cytoplasm, but a dense, periodic acid-Schiff (PAS)–positive, luminal secretory product is common. Hyperplasia typically is an incidental finding (McCluggage, 2000; Zaino, 2000, 2002; NHSCSP, 1999).

2.5.4 Ancillary techniques for distinction of pre-invasive lesions from benign mimics

Although the histological features of cervical glandular intraepithelial neoplasia are well described, but a wide variety of benign endocervical glandular lesions may be confused with CGIN and even invasive cervical adenocarcinoma. Many of these benign mimics are rare and in everyday practice the lesions most likely to be confused with CGIN are tuboendometrial metaplasia (TEM) and endometriosis. TEM is extremely common in the cervix, especially after loop or cone biopsy or some other operative procedure. The presence of cilia is a useful diagnostic clue to TEM, but these may be absent or inconspicuous, especially in cases showing endometrioid differentiation. Moreover, cervical TEM, especially when associated with a previous operative procedure, may have an altered stroma, raising the possibility of a desmoplastic reaction. Endometriosis within the superficial cervix may also cause diagnostic problems, especially when the characteristic stroma is inconspicuous. Fibrosis, caused by previous episodes of hemorrhage, may result in consideration of a desmoplastic stromal reaction (McCluggage, 2003).

While tuboendometrial metaplasia and endometriosis are especially likely to be misdiagnosed as cervical glandular intraepithelial neoplasia, microglandular hyperplasia (MGH) is more likely to be mistaken for an invasive adenocarcinoma, usually clear cell adenocarcinoma. A diagnosis of low-grade CGIN (LCGIN) especially is poorly reproducible and, in many institutions, this diagnosis is rarely, if ever, made in the absence of high-grade cervical glandular intraepithelial neoplasia (Cameron et al., 2002).

Microglandular hyperplasia is also common within the cervix. Although most cases are easily recognized, atypical features may be found, including the presence of signet ring cells, stromal hyalinization, or a lace-like growth pattern. These features may cause confusion with invasive cervical adenocarcinoma, especially of the clear cell type (McCluggage, 2003).

Immunohistochemical staining using a panel of antibodies, namely- MIB1, bcl2, and p16 - may be extremely useful in problematic cases in distinguishing these benign mimics from high grade CGIN (HCGIN) or invasive adenocarcinoma; although it has been emphasized that careful morphological examination is the mainstay of diagnosis (Cameron et al., 2002). The proliferation marker MIB1, which reacts with the Ki-67 antigen, has been shown to be a useful adjunct to histology in distinguishing HCGIN from benign mimics. A proliferation index of > 30% is generally indicative of HCGIN, whereas most cases of TEM, endometriosis, and MGH exhibit a proliferation index of < 10%. However, there may be some overlap, with occasional cases of HCGIN also exhibiting a proliferation index of < 10%. In addition, in some studies occasional benign lesions have exhibited a proliferation index of up to 50% (Nucci, 2002; Ostör et al., 2000).

In general, however, there are great differences in the MIB1 index between TEM, endometriosis, and HCGIN. Characteristically, many positive nuclei are present in HCGIN, with only scattered immunoreactivity in benign lesions. Immunohistochemical staining for bcl2 may also be useful in distinguishing TEM and endometriosis from HCGIN. Some studies have shown that cervical TEM and endometriosis (but not MGH) show consistent cytoplasmic expression of bcl2 (Cameron et al., 2002; McCluggage, 2002, 1997). Most cases of CGIN are negative. Why cervical TEM and endometriosis should exhibit positive staining for bcl2 is not certain but interestingly there is strong positive staining of normal fallopian tube epithelial cells and of proliferative endometrium with antibodies to bcl2. Of course, TEM and endometriosis are morphologically similar to normal fallopian tube and normal proliferative endometrium, respectively.

Also CD10 is a useful marker for confirming the presence of endometrial stroma and in establishing a diagnosis of endometriosis; however, this is of limited value in the cervix since a rim of CD10 reactive stromal cells surrounds normal endocervical glands (McCluggage, 2003).

In the distinction of benign mimics from HCGIN, p16 staining may also be of value. Some studies have shown overexpression of p16 in high grade cervical squamous intraepithelial lesions and in low grade lesions associated with high risk Human Papilloma Virus (HPV) types, p16 overexpression seems to be related to the presence of high risk HPV types (McCluggage et al., 2003; Pavlakis et al.,2006; Li et al.,2007; Riethdorf et al.,2002). Cameron et al. have founded a consistent positive staining of HCGIN (involving 100% of cells) with antibodies to p16. In contrast, cells of MGH were negative (Cameron et al., 2002). Staining of TEM and endometriosis was common but this was always focal and completely different to the pattern of immunoreactivity found in HCGIN. Thus, the combination of p16, MIB1, and bcl2 may be extremely useful in separating these benign mimics from HCGIN (McCluggage, 2003; Scheiden et al., 2004).

The diffuse distribution of p16 immunostaining in HPV 16/18 positive glandular neoplasms support a strong association with HPV infection and indicates that this biomarker mainly discriminate AIS from benign mimics (Riethdorf et al. , 2002). The situation with LCGIN has not been well studied and further work is necessary to ascertain whether these antibodies are of value in the separation of LCGIN from benign mimics. It is stressed that, in all cases, these antibodies are only of ancillary use and that careful morphological examination remains the cornerstone of diagnosis.

Immunohistochemical staining with carcinoembryonic antigen (CEA) has been reported to be of value in the separation of neoplastic endocervical glandular lesions and benign mimics (McCluggage, 2003). Diffuse cytoplasmic staining is usually present in neoplastic but not in benign lesions. However, as minimal deviation adenocarcinoma (MDA) is the neoplastic lesion most likely to be confused with benign lesions and as cytoplasmic staining with CEA may be focal and may not be present on a small biopsy, the value is limited. Conversely, normal endocervical epithelium may show luminal CEA positivity and some benign lesions, especially microglandular adenosis, may show cytoplasmic positivity, usually confined to areas of immature squamous metaplasia or reserve cell hyperplasia (McCluggage, 2003).

Other studies have found that a combination of CEA, MIB1, and p53 staining is useful in discriminating between benign and malignant endocervical glandular lesions (McCluggage, 2003; Pavlakis et al. , 2006). Polacarz et al. have shown *myc* immunostaining seemed to be a powerful discriminator between normal cervical glandular epithelium and epithelium show in intraepithelial changes or overt malignant changes. Apical cytoplasmic *myc* localisation thus seemed to be specific for CGIN and invasive adenocarcinoma of the cervix (Polacarz et al., 1991).

Other studies have evaluated the use of silver stained nucleolar organiser regions (AgNORs) in the separation of high grade CGIN and adenocarcinoma from benign histological mimics. In one study, significant differences in AgNOR counts were found between microglandular hyperplasia and HCGIN (McCluggage, 2000). However, the counting of AgNORs is laborious and time-consuming and is probably of less value than the use of proliferation markers.

In a recent study by Li et al.,their findings demonstrate significant expression of insulin-like growth factor-II mRNA-binding protein 3 (IMP3)and p16INK4a in adenocarcinoma in situ as compared to benign endocervical glands, suggesting that expression of these biomarkers may be helpful in the distinction of adenocarcinoma in situ from benign endocervical glands, particularly in difficult borderline cases (Li et al.,2007).

Findings of Little et al. study demonstrate that cyclin D1 can be included within an immunohistochemical panel to aid in the distinction between reactive cervical glandular lesions and adenocarcinoma in situ. The localized distribution of staining within invasive lesions suggests that cyclin D1 up-regulation has a specific role during the progression of some endocervical adenocarcinomas (Little & Stewart, 2010).

As result, immunohistochemical staining using a panel of antibodies may be very practical in problematic cases in distinguishing these benign mimics from high grade CGIN (HCGIN) or invasive adenocarcinoma; although it has been emphasized that careful morphological examination is the basis of diagnosis.

2.5.5 Approach to the diagnosis of cervical glandular intraepithelial neoplasia (CGIN) and pathological reporting

The approach to the diagnosis of CGIN is outlined below and is based on our experience and review of published article.

1. Generally speaking, the frequency of CGIN is low, and pathologists may rarely encounter to such a lesion in daily practice. Therefore in every case of cervical biopsy, it is rational to be aware of CGIN and its mimics and consider them in differential diagnosis.
2. Combination of invasive and pre-invasive lesions of squamous and glandular epithelium is a common event which has been reported in 30% to 70% of CGIN. Usually in low power examination, changes in a stratified squamous epithelium are more eye-catching and one may missed the concomitant glandular lesion. To prevent such a pitfall, we recommend to carefully examining the glandular epithelium architecturally and cytologically with low and high power field microscopy, especially when the lining of the canal and the glandular one had been replaced by a darker epithelium in each cervical specimen (Gloor & Ruzicka, 1982).
3. The next step is attention to any change in architectural pattern of endocervical glandular epithelium, including glandular branching, budding, crowding, infolding, villoglandular and cribriform, which is easily recognized even in low power microscopic examination. It is essential to emphasize that normal cleft and glands of endocervical epithelium can be variable in size and shape and may be mistaken for CGIN yet minimal deviation of adenocarcinoma. However comparison of the suspicious glands with unquestionably benign ones in the vicinity may provide guidance and attention to the following points should help to exclude CGIN or carcinoma: absence of cytologic atypia, desmoplastic response and marked variation in size and shape of endocervical glands. However, regardless of presence, these criteria have not solved the difficulty in diagnosis, and this may require the examination of additional tissue (e.g. Cervical cone).
4. Even though some cytological features including stratification, mucin depletion and abrupt junction between normal and abnormal columnar epithelium, can be recognized in low power microscopic examination, emergence of a darker epithelium which indicate replacement of normal epithelium by stratified epithelium may be helpful.
5. As mentioned before, architectural pattern may be associated with benign conditions (cervicitis, tubal metaplasia, endometriosis, tunnel clusters and etc) or invasive adenocarcinoma and then must be combined with cytological features. The cytological features are nuclear changes, apoptotic bodies, mitotic figures and intestinal metaplasia can be evaluated exactly by x10, x 40 microscopic power examinations. Because many of cytological features may be associated with benign reactive changes or metaplastic condition; pathologist must be aware and combined cytological and architectural features for final diagnosis. High N/C ratio of the columnar epithelium in some metaplastic conditions can mimic endocervical or tubal type of AIS. Increased mitotic index (MI) especially atypical mitosis is a clue in the diagnosis of CGIN. The average MI of CGIN is intermediate between benign condition and invasive adenocarcinoma (Moritani et al.,

2002). Although mitosis is uncommon in benign condition, it is occasionally seen in endometriosis, estrogen consumption, and in the repair process (NHSCSP, 1999).

6. Apoptotic bodies are useful in establishing a diagnosis of CGIN, although they may not be prominent in all cases. Apoptotic body is a constant feature of HCGIN and the increase number of apoptotic bodies was significantly higher than in nonspecific endocervical glandular lesions (Moritani et al., 2002).

7. Most cases of CGIN are of usual endocervical type. However, other rare variants have been described. An endometrioid variant of CGIN has been reported. However, this is rare (if it occurs at all) and most cases diagnosed as such are probably cases of usual endocervical-type CGIN with scant intracytoplasmic mucin. An intestinal variant of HCGIN exists and is not uncommon. This is characterized by the presence of goblet cells and less commonly paneth or neuroendocrine cells (McCluggage, 2003). These microscopic features are along with Gloor study that was named CGIN type B alongside all other mentioned above features that described as CGIN type A (Pirog et al., 2000). It is doubtful whether intestinal differentiation in endocervical glands ever occurs without coexistent CGIN or invasive adenocarcinoma. Benign intestinal metaplasia involving endocervical glands has been described, but it is probably an extremely rare phenomenon, if it occurs at all, and the presence of goblet cells almost always indicates CGIN (Ioffe, 2003).

In regard to above mentioned approach the following points are emphasized in histological reporting of CGIN because these factors would influence the management:

1. Lesion location: exocervical, endocervical or both
2. Tridimensional lesion geometry: linear length of lesion and underlying crypt involvement depth. Clearly, there is no consensus in the acceptable depth of involvement. Ostor study revealed the depth of crypt involvement (measured from the surface) varied from 1.5 to 4 mm with an average of 2.6mm.The length of extent as measured horizontally in single section ranged from 0.5 to 30mm, with an average of 7mm.The width of the lesion (as determined from the number of blocks involved) ranged from 0.5 mm to 25 mm, with an average of 12 mm. From these parameters hypothetical tumor volumes could be calculated, the smallest being 0.25mm³, and the largest 1,500 mm³.The average tumor volume was 313 mm³ (Ostör et al.,1984).
3. Potentional for AIS to be buried under metaplstic or dysplastic epithelium
4. Presence of squamous component (Colgan & Lickrish, 1990)
5. Possibly multifocal lesions and skip lesions
6. Possibly multicentric lesions (more than one quadrant involvement) or circumferential extent (Sheets, 2002)
7. Specimen margin status (post excision)

2.6 Biological behavior/management

It is beyond the scope of this chapter to discuss in detail the management of CGIN but it is clear that in those who wish to preserve their fertility (HCGIN and early invasive adenocarcinoma often occur in young women), local excision with careful pathological examination and free margins combined with close cytological follow up may be used for treatment. After completion of childbearing , hysterectomy is necessary because of the paucity of data concerning the long-term natural history of CGIN (Ostör, 2000; Sheets, 2002; Zhao et al., 2009).

3. Conclusion

There is a real increase in the incidence of malignant and premalignant endocervical glandular lesions, which are thus arrogant increasing importance in diagnostic surgical pathology but the frequency of CGIN is low, and pathologists may rarely encounter to such a lesion in daily practice. Therefore in every case of cervical biopsy, it is rational to be aware of CGIN and its mimics and consider them in differential diagnosis. In most, but not all, cases CGIN occurs close to the transformation zone and there is often an associated squamous intraepithelial lesion. CGIN can be confused with a wide variety of benign endocervical glandular lesions and even invasive cervical adenocarcinoma. CGIN should be classified as low grade or high grade CGIN. High grade CGIN is alternatively known as AIS and Low grade CGIN is alternatively known as glandular dysplasia. High grade CGIN is a vigorous diagnosis but distinction from early invasive adenocarcinoma may be difficult and Low grade CGIN may be underdiagnosed by pathologists. A combination of architectural and cytological features is necessary for diagnosis of CGIN. Immunohistochemical staining using a panel of antibodies may be useful in difficult cases in distinctive benign mimics from high grade CGIN or invasive carcinoma , although it is stressed that careful morphological examination is the basis of diagnosis.

4. Acknowledgment

The authors would like to thank Dr. Sanaz Sanii for preparing the proposal text.

5. Common abbreviations

AIS: Adenocarcinoma in situ
CGIN: Cervical glandular intraepithelial neoplasia
CIN: Cervical intraepithelial neoplasia
HCGIN: High grade CGIN
LCGIN: Low grade CGIN
MGH: Microglandular hyperplasia
MIA: Microinvasive adenocarcinoma
SCC: Squamous cell carcinoma
TBS: The Bethesda System
TEM: Tuboendometrioid metaplasia

6. Referrences

Bekkers, R. L., Bulten, J., Wiersma-van Tilburg, A., Mravunac, M., Schijf, C. P., Massuger, L. F., et al. (2003). Coexisting high-grade glandular and squamous cervical lesions and human papillomavirus infections. *Br J Cancer, 89*(5), 886-890.

Brown, L. J., & Wells, M. (1986). Cervical glandular atypia associated with squamous intraepithelial neoplasia: a premalignant lesion? *J Clin Pathol, 39*(1), 22-28.

Bulk, S., Berkhof, J., Bulkmans, N. W., Zielinski, G. D., Rozendaal, L., van Kemenade, F. J., et al. (2006). Preferential risk of HPV16 for squamous cell carcinoma and of HPV18 for adenocarcinoma of the cervix compared to women with normal cytology in The Netherlands. *Br J Cancer, 94*(1), 171-175.

Cameron, R. I., Maxwell, P., Jenkins, D., & McCluggage, W. G. (2002). Immunohistochemical staining with MIB1, bcl2 and p16 assists in the distinction of cervical glandular intraepithelial neoplasia from tubo-endometrial metaplasia, endometriosis and microglandular hyperplasia. *Histopathology, 41*(4), 313-321.

Campion M.J.(2010). Preinvasive disease, in: *Berek & Hacker Gynecologic Oncology,* Berek J.S. & Hacker N.F, (Ed.), pp. 3132-414, LWW, ISBN 0781795125, Philadelphia ,USA.

Christopherson, W. M., Nealon, N., & Gray, L. A., Sr. (1979). Noninvasive precursor lesions of adenocarcinoma and mixed adenosquamous carcinoma of the cervix uteri. *Cancer, 44*(3), 975-983.

Colgan, T. J., & Lickrish, G. M. (1990). The topography and invasive potential of cervical adenocarcinoma in situ, with and without associated squamous dysplasia. *Gynecol Oncol, 36*(2), 246-249.

Covell J.L.,Wilbur D.C.,Guidos B.,Lee K.R.,Chieng D.C.,Mody D.R. (2003). Epithelial abnormalities: glandular. In: *The Bethesda System for Reporting Cervical Cytology,* Solomon D., Nayar R.,(Eds). pp. 123-141. Spriger. ISBN 0387403582, New York, USA.

Friedell, G. H., & Mc, K. D. (1953). Adenocarcinoma in situ of the endocervix. *Cancer, 6*(5), 887-897.

Gloor, E., & Ruzicka, J. (1982). Morphology of adenocarcinoma in situ of the uterine cervix: a study of 14 cases. *Cancer, 49*(2), 294-302.

Gloor, E., & Hurlimann, J. (1986). Cervical intraepithelial glandular neoplasia (adenocarcinoma in situ and glandular dysplasia). A correlative study of 23 cases with histologic grading, histochemical analysis of mucins, and immunohistochemical determination of the affinity for four lectins. *Cancer, 58*(6), 1272-1280.

Histopathology Reporting in Cervical Screening.(1999). *NHSCSP Publication.* No 10. April 1999.

Hopkins, M. P., & Morley, G. W. (1991). A comparison of adenocarcinoma and squamous cell carcinoma of the cervix. *Obstet Gynecol, 77*(6), 912-917.

Ioffe, O. B., Sagae, S., Moritani, S., Dahmoush, L., Chen, T. T., & Silverberg, S. G. (2003). Symposium part 3: Should pathologists diagnose endocervical preneoplastic lesions "less than" adenocarcinoma in situ?: Point. *Int J Gynecol Pathol, 22*(1), 18-21.

Kurian, K., & al-Nafussi, A. (1999). Relation of cervical glandular intraepithelial neoplasia to microinvasive and invasive adenocarcinoma of the uterine cervix: a study of 121 cases. *J Clin Pathol, 52*(2), 112-117.

Kurman RJ, Ronnett BM, Sherman ME, Wilkinson EJ. (2010).*Tumours of the cervix, vagina and vulva. Atlas of tumor pathology,* 4th series, Fascicle 13 ,American Registry of Pathology,1-933477-11-3. Washington,DC

Leminen, A., Paavonen, J., Forss, M., Wahlstrom, T., & Vesterinen, E. (1990). Adenocarcinoma of the uterine cervix. *Cancer, 65*(1), 53-59.

Li, C., Rock, K. L., Woda, B. A., Jiang, Z., Fraire, A. E., & Dresser, K. (2007). IMP3 is a novel biomarker for adenocarcinoma in situ of the uterine cervix: an immunohistochemical study in comparison with p16(INK4a) expression. *Mod Pathol, 20*(2), 242-247.

Liang, J., Mittal, K. R., Wei, J. J., Yee, H., Chiriboga, L., & Shukla, P. (2007). Utility of p16INK4a, CEA, Ki67, P53 and ER/PR in the differential diagnosis of benign, premalignant, and malignant glandular lesions of the uterine cervix and their relationship with Silverberg scoring system for endocervical glandular lesions. *Int J Gynecol Pathol, 26*(1), 71-75.

Little, L., & Stewart, C. J. Cyclin D1 immunoreactivity in normal endocervix and diagnostic value in reactive and neoplastic endocervical lesions. *Mod Pathol, 23*(4), 611-618.

Madeleine, M. M., Daling, J. R., Schwartz, S. M., Shera, K., McKnight, B., Carter, J. J., et al. (2001). Human papillomavirus and long-term oral contraceptive use increase the risk of adenocarcinoma in situ of the cervix. *Cancer Epidemiol Biomarkers Prev, 10*(3), 171-177.

Marques, J. P., Costa, L. B., Pinto, A. P., Lima, A. F., Duarte, M. E., Barbosa, A. P., et al. Atypical glandular cells and cervical cancer: systematic review. *Rev Assoc Med Bras, 57*(2), 234-238.

McCluggage, G., McBride, H., Maxwell, P., & Bharucha, H. (1997). Immunohistochemical detection of p53 and bcl-2 proteins in neoplastic and non-neoplastic endocervical glandular lesions. *Int J Gynecol Pathol, 16*(1), 22-27.

McCluggage, W. G., & Maxwell, P. (2002). bcl-2 and p21 immunostaining of cervical tubo-endometrial metaplasia. *Histopathology, 40*(1), 107-108.

McCluggage, W. G., Oliva, E., Herrington, C. S., McBride, H., & Young, R. H. (2003). CD10 and calretinin staining of endocervical glandular lesions, endocervical stroma and endometrioid adenocarcinomas of the uterine corpus: CD10 positivity is characteristic of, but not specific for, mesonephric lesions and is not specific for endometrial stroma. *Histopathology, 43*(2), 144-150.

McCluggage, W. G. (2003). Endocervical glandular lesions: controversial aspects and ancillary techniques. *J Clin Pathol, 56*(3), 164-173.

McCluggage, W. G. (2007). Immunohistochemistry as a diagnostic aid in cervical pathology. *Pathology, 39*(1), 97-111.

Mood, N. I., Eftekhar, Z., Haratian, A., Saeedi, L., Rahimi-Moghaddam, P., & Yarandi, F. (2006). A cytohistologic study of atypical glandular cells detected in cervical smears during cervical screening tests in Iran. *Int J Gynecol Cancer, 16*(1), 257-261.

Moritani, S., Ioffe, O. B., Sagae, S., Dahmoush, L., Silverberg, S. G., & Hattori, T. (2002). Mitotic activity and apoptosis in endocervical glandular lesions. *Int J Gynecol Pathol, 21*(2), 125-133.

Nasu, I., Meurer, W., & Fu, Y. S. (1993). Endocervical glandular atypia and adenocarcinoma: a correlation of cytology and histology. *Int J Gynecol Pathol, 12*(3), 208-218.

Nucci, M. R. (2002). Symposium part III: tumor-like glandular lesions of the uterine cervix. *Int J Gynecol Pathol, 21*(4), 347-359.

Ostor, A. G., Pagano, R., Davoren, R. A., Fortune, D. W., Chanen, W., & Rome, R. (1984). Adenocarcinoma in situ of the cervix. *Int J Gynecol Pathol, 3*(2), 179-190.

Ostor, A. G. (2000). Early invasive adenocarcinoma of the uterine cervix. *Int J Gynecol Pathol, 19*(1), 29-38.

Parazzini, F., La Vecchia, C., Negri, E., Fasoli, M., & Cecchetti, G. (1988). Risk factors for adenocarcinoma of the cervix: a case-control study. *Br J Cancer, 57*(2), 201-204.

Pavlakis, K., Messini, I., Athanassiadou, S., Kyrodimou, E., Pandazopoulou, A., Vrekoussis, T., et al. (2006). Endocervical glandular lesions: a diagnostic approach combining a semi-quantitative scoring method to the expression of CEA, MIB-1 and p16. *Gynecol Oncol, 103*(3), 971-976.

Pecorelli, S. (2009). Revised FIGO staging for carcinoma of the vulva, cervix, and endometrium. *Int J Gynaecol Obstet, 105*(2), 103-104.

Pirog, E. C., Kleter, B., Olgac, S., Bobkiewicz, P., Lindeman, J., Quint, W. G., et al. (2000). Prevalence of human papillomavirus DNA in different histological subtypes of cervical adenocarcinoma. *Am J Pathol, 157*(4), 1055-1062.

Pirog, E. C., Isacson, C., Szabolcs, M. J., Kleter, B., Quint, W., & Richart, R. M. (2002). Proliferative activity of benign and neoplastic endocervical epithelium and correlation with HPV DNA detection. *Int J Gynecol Pathol, 21*(1), 22-26.

Polacarz, S. V., Darne, J., Sheridan, E. G., Ginsberg, R., & Sharp, F. (1991). Endocervical carcinoma and precursor lesions: c-myc expression and the demonstration of field changes. *J Clin Pathol, 44*(11), 896-899.

Riethdorf, L., Riethdorf, S., Lee, K. R., Cviko, A., Loning, T., & Crum, C. P. (2002). Human papillomaviruses, expression of p16, and early endocervical glandular neoplasia. *Hum Pathol, 33*(9), 899-904.

Scheiden, R., Wagener, C., Knolle, U., Dippel, W., & Capesius, C. (2004). Atypical glandular cells in conventional cervical smears: incidence and follow-up. *BMC Cancer, 4*, 37.

Sheets, E. E. (2002). Management of adenocarcinoma in situ, micro-invasive, and early stage adenocarcinoma of the cervix. *Curr Opin Obstet Gynecol, 14*(1), 53-57.

Smedts, F., Ramaekers, F. C., & Hopman, A. H. The two faces of cervical adenocarcinoma in situ. *Int J Gynecol Pathol, 29*(4), 378-385.

Solomon, D., Frable, W. J., Vooijs, G. P., Wilbur, D. C., Amma, N. S., Collins, R. J., et al. (1998). ASCUS and AGUS criteria. International Academy of Cytology Task Force summary. Diagnostic Cytology Towards the 21st Century: An International Expert Conference and Tutorial. *Acta Cytol, 42*(1), 16-24.

Tam, K. F., Cheung, A. N., Liu, K. L., Ng, T. Y., Pun, T. C., Chan, Y. M., et al. (2003). A retrospective review on atypical glandular cells of undetermined significance (AGUS) using the Bethesda 2001 classification. *Gynecol Oncol, 91*(3), 603-607.

Ursin, G., Peters, R. K., Henderson, B. E., d'Ablaing, G., 3rd, Monroe, K. R., & Pike, M. C. (1994). Oral contraceptive use and adenocarcinoma of cervix. *Lancet, 344*(8934), 1390-1394.

van Aspert-van Erp, A. J., Smedts, F. M., & Vooijs, G. P. (2004). Severe cervical glandular cell lesions with coexisting squamous cell lesions. *Cancer, 102*(4), 218-227.

Wang, S. S., Sherman, M. E., Hildesheim, A., Lacey, J. V., Jr., & Devesa, S. (2004). Cervical adenocarcinoma and squamous cell carcinoma incidence trends among white women and black women in the United States for 1976-2000. *Cancer, 100*(5), 1035-1044.

Wells, M., & Brown, L. J. (2002). Symposium part IV: investigative approaches to endocervical pathology. *Int J Gynecol Pathol, 21*(4), 360-367.

Wright V.C. (2002). Colposcopic features of cervical adenocarcinoma in situ and adenocarcinoma and management of preinvasive disease. In: *Colposcopy Principles and Practice*. Apgar B.S., Brotzaman G.L. & Spitzer M. (Ed.).pp.301-303. Saunders. ISBN 1416034056. Philadelphia, USA

Zaino, R. J. (2000). Glandular lesions of the uterine cervix. *Mod Pathol, 13*(3), 261-274.

Zaino, R. J. (2002). Symposium part I: adenocarcinoma in situ, glandular dysplasia, and early invasive adenocarcinoma of the uterine cervix. *Int J Gynecol Pathol, 21*(4), 314-326.

Zhao, C., Florea, A., Onisko, A., & Austin, R. M. (2009). Histologic follow-up results in 662 patients with Pap test findings of atypical glandular cells: results from a large academic womens hospital laboratory employing sensitive screening methods. *Gynecol Oncol, 114*(3), 383-389.

Zielinski, G. D., Snijders, P. J., Rozendaal, L., Daalmeijer, N. F., Risse, E. K., Voorhorst, F. J., et al. (2003). The presence of high-risk HPV combined with specific p53 and p16INK4a expression patterns points to high-risk HPV as the main causative agent for adenocarcinoma in situ and adenocarcinoma of the cervix. *J Pathol, 201*(4), 535-543.

3

Cervical Intraepithelial Neoplasia (CIN) (Squamous Dysplasia)

Oguntayo Olanrewaju Adekunle
*Department of Obstetric and Gynaecology, Ahmadu Bello
University Teaching Hospital, Zaria Kaduna State
Nigeria*

1. Introduction

Cervical intraepithelial neoplasia (CIN) is a premalignant cervical disease that is also called cervical dysplasia or **cervical interstitial neoplasia** or cervical squamous intraepithelial lesions (CSIL).

The nomenclature in use in the past was mild, moderate, and severe dysplasia, these were the terms used to describe premalignant squamous cervical cellular changes. Although still in use by some, it has generally been replaced by the term Cervical Intraepithelia Neoplasia(CIN), which is used to describe histologic changes on the uterine cervix. The trend is now tending towards the use of Squamous Intraepithelia Lesions(SIL).

1.1 Definition

What is cervical intraepithelial Neoplasia-? It is a potentially premalignant transformation and abnormal growth (dysplasia) of squamous cells on the surface of the cervix.[Kumar etal 2007] CIN is not cancer, and is usually curable.[ACOG 2010] Most cases of CIN remain stable, or are eliminated by the host's immune system without intervention. However a small percentage of cases progress to become cervical cancer, usually cervical squamous cell carcinoma (SCC), if left untreated.[Agorastos et al 2005]

It can actually be defined as a spectrum of intraepithelial changes (dysplasia) with indistinct boundaries that begins with mild atypia and progresses through stages of more marked intraepithelial abnormalities to carcinoma in situ if untreated or managed. (UVA Health)

1.1.1 Dysplasia is a potentially reversible change characterized by an increase in mitotic rate, atypical cytologic features (size, shape, nuclear features) and abnormal organization (cellularity, differentiation, polarity) that fall short of invasive carcinoma (premalignant change).Dysplasia may progress to cancer and dysplastic changes may be found adjacent to foci of cancer.

1.2 Epidemiology

Population distribution of cervical intraepithelial neoplasia/dysplasia resembles the epidemiology of an infectious disease that is sexually transmitted. Multiple male sexual partners, early age at first sexual intercourse and male partner with multiple previous/current female sexual partners are very important risk factors.

1.3 Incidence

The estimated annual incidence in the United States of CIN among women who undergo cervical cancer screening is 4 percent for CIN 1 and 5 percent for CIN 2,3 [Agorastos et al 2005]. High grade lesions are typically diagnosed in women 25 to 35 years of age, while invasive cancer is more commonly diagnosed after the age of 40, typically 8 to 13 years after a diagnosis of a high grade lesion. Between 250,000 and 1 million American women are diagnosed with CIN annually. Women can develop CIN at any age, however, women generally develop it between the ages of 25 to 35.[Kumar etal 2007]

In developing Nations like Nigeria the mean age for cervical intraepithelial neoplasia (CIN) was 37.6 years. CIN I accounted for 3.6%, CIN II 0.8% and CIN III was only 0.4%.The combined prevalence was 48 per 1000.The peculiarity of the developing nations result is the poor uptake or use of screening methods(Oguntayo & Samaila).

In view of the fact that CIN is a premalignant or precursor of cervical cancer it is pertinent to briefly see the incidence and prevalence of this disease condition.

Cervical cancer is second only to breast cancer in its incidence world wide. Cancer registry data shows that that there are approximately 400,000 new cases of cervical cancer and 200,000 deaths from this disease every year (IARC 2001).

The incidence rate varies from country to country with eighty percent (80%) of the cases occurring in less developed countries. The reasons for this may lie in the socio economic conditions that prevail in these countries where facilities for family planning, obstetric and gynaecological health care are scarce and cervical screening programmes are virtually non existent. (IARC 2001).

The relative incidences of cervical intra-epithelial neoplasia (CIN) and invasive cervical cancer were studied in black and white patients at the academic hospitals of the University of the Orange Free State. A statistically high significant differences was found between black and white patients, with a higher incidence of invasive cervical cancer than stage III CIN (CIN III) in black patients and a higher incidence of CIN III than invasive cervical cancer in white patients (P=0,000092; 95% confidence interval -0,355 - -0,128). The time interval between the peak incidence of CIN III and that of invasive cervical cancer was found to be shorter in black than in white patients. These distressing findings emphasise the urgent need for a national cervical cytological screening programme to decrease the incidence of invasive cervical cancer (NEL1994)

2. Anatomy

2.1 Anatomy of the uterine cervix

The cervix is actually the lower, narrow portion of the uterus, connected to the uterine fundus by the uterine isthmus. Its name is derived from the Latin word for "neck." It is cylindrical or conical in shape. Its upper limit is considered to be the internal os, which is an anatomically and histologically ill-defined junction of the more muscular uterine body and the denser, more fibrous cervical stroma. The size and shape of the cervix varies widely with age, hormonal state, and parity. In parous women, the cervix is bulkier and the external os, or lowermost opening of the cervix into the vagina, appears wider and more slit-like and gaping than in nulliparous fig 2 women. Before childbearing, the external os is a small, circular opening at the center of the cervix fig 1. After the menopause it may narrow almost to a pin point fig 3. The portion of the cervix exterior to the external os is called the ectocervix. The passageway between the external os and the body of the uterus at the isthmus above is referred to as the endocervical canal. Its upper limit is the internal os.5

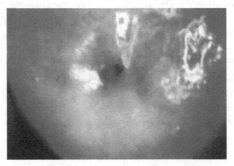

Fig. 1. The nulliparous cervix: note the small round os

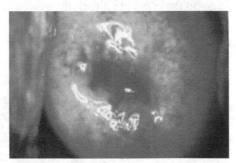

Fig. 2. Multiparous cervix

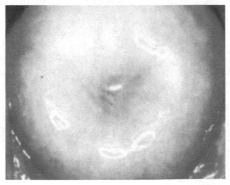

Fig. 3. Menopausal cervix

The canal itself shows a complex configuration of mucosal folds or plicae. These make cytologic screening and colposcopy of the endocervical tissues technically more difficult and less reliable than for the smoother and more accessible squamous epithelium of the ectocervix.(ASCCP 2011)

2.2 Embryology of the uterine cervix

Two paramesonephric ducts form from coelomic epithelium extending beside the mesonephric ducts. In the absence of Mullerian Inhibitory Factor these ducts proliferate and grow extending from the vaginal plate on the wall of the urogenital sinus to lie beside the developing ovary. The paired ducts begin to fuse from the vaginal plate end, forming the primordial body of the uterus and the unfused lateral arms form the uterine tubes.see fig 4

The picture bellow is the summary of the embryonic development of the uterine cervix and the second is showing the infantine uterus as it appears and this is an evidence that there are significant changes that do occur as the girl child grows fig 5.The main clinical reference to this is basically in the epithelia changes between pre pubertal and post pubertal period.The epithelia lining of the cervical canal (endocervix) is the columnar epithelium while that of external cervix (endocervix) is squamous epithelium. The squamo-columnar junction is located at the point where the squamous epithelium and the columnar epithelium meet. The location varies throughout a woman's life due to the process of metaplastic changes in the cervical epithelium which occur after puberty and in pregnancy.(Mark 2010)

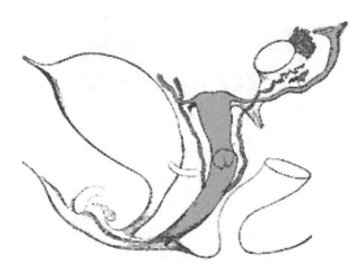

Fig. 4. Embryological Origin of the Uterus

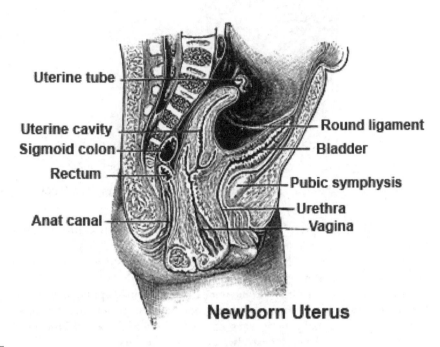

Uterine tube

Uterine cavity

Sigmoid colon

Rectum

Anat canal

Round ligament

Bladder

Pubic symphysis

Urethra

Vagina

Newborn Uterus

Fig. 5.

2.3 Cervix-normal histology

Most of the cervix is composed of fibromuscular tissue. The Epithelium is either squamous or columnar.

The endocervix is lined by columnar epithelium that secretes mucus this epithelium has complex infoldings that resemble glands or clefts on cross section and the mucosa rests on inconspicuous layer of reserve cells.

The ectocervix (exocervix)is covered by nonkeratinizing, stratified squamous epithelium, either native or metaplastic; has basal, midzone and superficial layers. A fter menopause and in prepubertal girls the superficial layer becomes atrophic with mainly basal and parabasal cells with high nucleo-cytoplasmic ratio that resembles dysplasia.

2.3.1 Squamocolumnar junction: where squamous and glandular (columnar) epithelium meets this a major land mark in cervical dyplasia,it is usually in exocervix.The nearby reserve cells are involved in squamous metaplasia, dysplasia and carcinoma.

2.3.2 Transformation zone: also called ectropion, between original squamocolumnar junction and border of **metaplastic squamous epithelium;** epidermalization and squamous differentiation of reserve cells transform this area to squamous epithelium; site of squamous cell carcinomas and dysplasia.

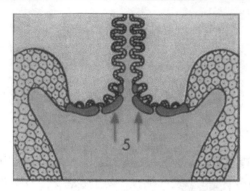

Metaplastic change of endocervical epithelium in the transformation zone

In the cervix a lot of metaplasia takes place which was what encouraged a lot of study to be conducted.

- Metaplasia is the name given to the process by which one fully differentiated type of epithelium changes into another.
- It is usually an adaptive change which occurs in reaction to longstanding (chronic) irritation of any kind, or in response to hormonal stimuli.
- Metaplastic change is reversible and theoretically transformed epithelium should revert to its original form after the stimulus is removed but this does not always happen.
- Metaplasia occurs at many body sites eg gastric mucosa, bladder,bronchi etc. The metaplastic process has been extensively studied in the cervix.

2.4 Physiology of metaplastic changes on the cervix

The metaplastic changes seen are related to the following:

Pre-Puberty, Post puberty, Pregnancy and Menopause.

Pre-puberty-From birth until puberty the endocervical epithelium is composed of columnar epithelium and the ectocervix of native squamous epithelium. The interface between the two is termed the original squamocolumnar junction.

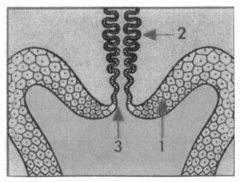

Squamocolumnar junction prior to puberty.

Puberty-During puberty and at the first pregnancy the cervix increases in volume in response to hormonal changes. The endocervical epithelium everts onto the ectocervix (portio vaginalis) exposing it to the acid pH of the vagina. This provides a stimulus for metaplastic change of the columnar epithelium.

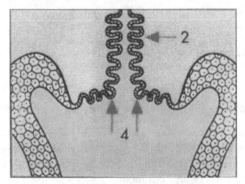

Eversion of the endocervical epithelium at puberty and first pregnancy

Menopause-The process of metaplasia is a patchy one: It starts initially in the crypts and at the tips of the endocervical villae which gradually fuse. Eventually the whole of the everted endocervical epithelium may be replaced by squamous epithelium.

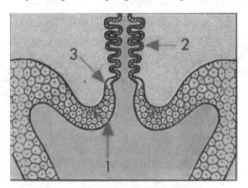

Relocation of SCJ in the endocervical canal after the menopause

Key:

- 1: native squamous epithelium
- 2: columnar epithelium of endocervix
- 3: squamocolumnar junction (SCJ)
- 4: Eversion of endocervical epithelium
- 5: Metaplastic change in transformation zone (Eurocytology 2011)

The Clinical significance of squamous metaplasia in the cervix is that, this area of the cervical epithelium has under gone metaplasia(Transformation zone) and all the immature metaplastic are susceptible to carcinogens.In view of the afore mentioned it is not surprising that most cervical cancers arise here.

2.4.1 Basal cells (reserve cells): cuboidal to low columnar with scant cytoplasm and round/oval nuclei; acquire eosinophilic cytoplasm as they mature; positive for low molecular weight keratin and estrogen receptor; negative for high molecular weight keratin and involucrin

2.4.2 Suprabasal cells: have variable amount of glycogen, detectable with Lugol/Schiller's test (application of iodine)

2.4.3 Glandular epithelium: positive for estrogen receptor.

3. Aetiology

3.1 Association of human papillomaviruses and cervical Intraepithelia neoplasia

Human papillomaviruses (HPV) are members of a family of viruses known as the Papovaviruses fig 5. They are epitheliotropic viruses which promote cell proliferation which results in the development of benign papillomatous lesions of the genital tract upper respiratory tract, digestive tracts and cutaneous lesions of the skin. More than 70 distinct HPV types have been identified as a result of molecular hybridisation of DNA extracted from condylomata or warty lesions from a variety of sites. Each virus type has a very restricted site of infection and viruses which occupy similar niches appear to be genetically related. Molecular hybridisation of anogenital warts and cervical biopsies have shown that about 30 of the 70 distinct types of HPV are confined to the female genital tract. (Eurocytology).

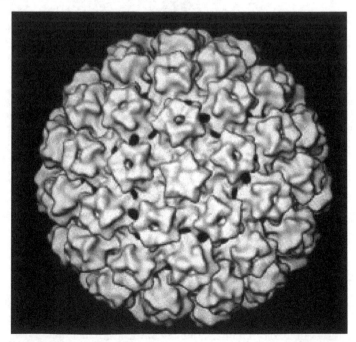

Fig. 6. Electronmicrograph of human papillomavirus (courtesy Eurocytology)

DNA analysis of anogenital warts,CIN and cervical cancerous tissue has shown that two groups of HPV can be identified in the female genital tract. One group of HPV is almost always associated with low grade CIN lesions and exophytic anogenital warts which have a *low risk* **of progressing to cervical cancer**. A second group of viruses is found most commonly in CIN2 and CIN3 which have a *high risk* **of developing into invasive cancer.**

3.1.1 HPV types found in the female genital tract

The major cause of CIN is chronic infection of the cervix with the sexually transmitted human papillomavirus (HPV), especially the high-risk HPV types 16 or 18(**viruses from the high risk group (HPV16 and HPV 18) have the ability to immortalise primary human keratinocytes** *ie* **extend their lifespan)** In comparison viruses from the low risk group (HPV-6 and HPV -11) do not extend the life span of transfected human cells which mature and die at the same rate as non infected cells fig 7. Similarly the low risk viruses perform poorly in experiments concerned with the malignant transformation of rodent cells in comparison to the high risk HPV types. Moreover, HPV-16 and HPV -18 infected human keratinocytes in raft culture (an organotypic culture medium) exhibit a differentiation pattern very similar to that seen in vivo in CIN.(Eurocytology). Over 100 types of HPV have been identified. About a dozen of these types appear to cause cervical dysplasia and may lead to the development of cervical cancer. Other types cause warts. Wikipedia,

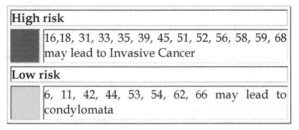

High risk	
	16,18, 31, 33, 35, 39, 45, 51, 52, 56, 58, 59, 68 may lead to Invasive Cancer
Low risk	
	6, 11, 42, 44, 53, 54, 62, 66 may lead to condylomata

Fig. 7.

3.2 The viral DNA Integration

The viral DNA Integration is a consistent finding in all cancers harbouring the *high risk* virus types HPV16 and HPV18 and provides the **strongest evidence** that HPV16 and HPV18 play an important role in the development of cervical cancer. **HPV DNA is present in 90% of all cervical invasive cancer.**

It is not sufficient to say that simple infection with high risk HPV or even integration of HPV 16 /18 into the host cell nucleus is enough for malignant transformation of the cervical epithelium.Obviously Infection of the genital tract with HPV 16 is relatively common whereas invasive cancer is rare; and integration has been detected in some cases of genital warts and CIN lesions. A number of associated-factors have been proposed such as impaired immune response, persistence of virus, smoking and administration of steroid hormones (as oral contraceptives). Other genetic events such as loss of tumour suppressor genes and the activation of oncogenes may also play a role . Mutations in *ras ,fos* and other oncogenes have been detected in cervical cancer cell lines but their role *in vivo* is still to be determined.The knowledge of HPV infection has made a remarkable improvement in the screening,diagnosis,treatment,prevention and prognosis of cancer of the cervix.

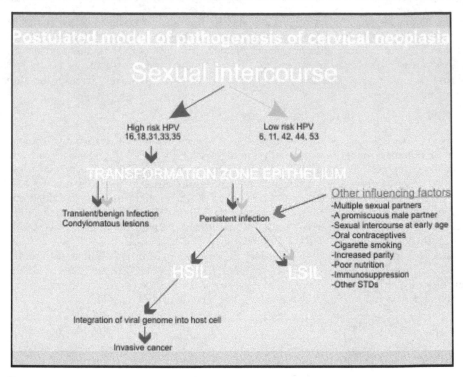

Fig. 8.

3.3 The host immune system and HPV

The host immunity plays a significant role in the control of this disease entity. The fact that HPV remains localised to cervix and vagina further indicates that local immune responses are sufficient in controlling and resolving HPV infection. Both cell mediated immunity and humoral immunity fig 8. Also immunosuppression has been implicated as an associated factor. The majority of infections are transient and not clinically evident with 70-90% of infections clearing within 12-30 months. This suggests that host immunity is generally able to clear HPV infection.

4. Histopathological features

4.1 Histo-pathological changes

Abnormal cellular proliferation, maturation and atypia characterize cervical intraepithelial neoplasia(CIN). Nuclear abnormality is the hallmark of CIN and includes hyperchromasia, pleomorphism, irregular borders, and abnormal chromatin distribution. These nuclear abnormalities persist throughout the epithelium irrespective of cytoplasmic maturation towards the surface. Mitotic rate is increased and abnormal mitotic figures may be seen.

Histologic grading of CIN is based on the proportion of the epithelium occupied by dysplastic cells. The epithelium is divided into thirds.

4.2 Grading

- **4.2.1 CIN 1** is considered a low grade lesion. It refers to mildly atypical cellular changes in the lower third (basal 1/3) of the epithelium (formerly called **mild dysplasia**/Abnormal cell growth). HPV viral cytopathic effect (koilocytotic atypia) is often present. This corresponds to infection with HPV, and typically will be cleared by immune response in a year or so, though can take several years to clear.
- **4.2.2 CIN 2** is considered a high grade lesion. It refers to moderately atypical cellular changes confined to the basal **two-thirds** of the epithelium (formerly called moderate dysplasia) with preservation of epithelial maturation.
- **4.2.3 CIN 3** is also considered a high grade lesion/Severe dysplasia. It refers to severely atypical cellular changes encompassing **greater than two-thirds** of the epithelial thickness, and includes full-thickness lesions (formerly called severe dysplasia or carcinoma in situ).

CIN 1 (mild dysplasia): Dysplastic cells occupy the lower third of the epithelium.fig 9

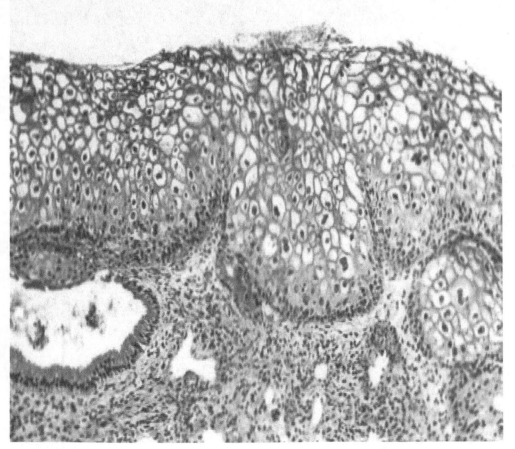

Fig. 9.

CIN 2 (moderate dysplasia): Dysplastic cells occupy up to the middle third of the epithelium. See fig 10.

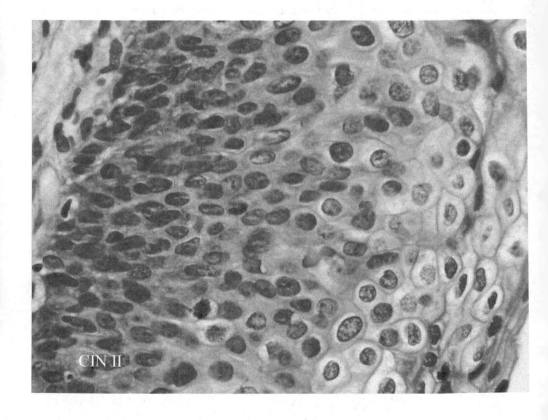

Fig. 10. CIN 2. Note superficial koilocytosis.

CIN 3 (severe dysplasia, carcinoma in situ): Dysplastic cells extend into the upper third and may occupy the full thickness of the epithelium.fig 11

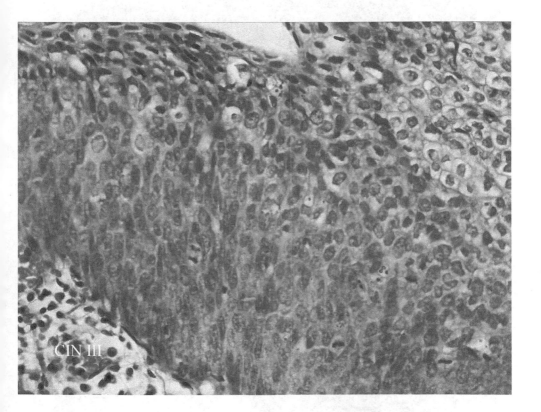

Fig. 11. CIN3. Note adjacent koilocytes (bottom right)

Cytologic grading of CIN also uses a three-tier system. However, the new Bethesda System for cytological diagnosis divides precursors of cervical squamous cell carcinoma into *low-grade squamous intraepithelial lesion* and *high-grade intra-epithelial lesion*.

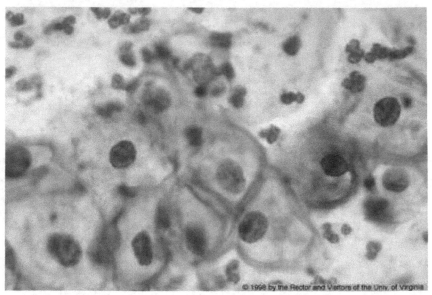

Fig. 12. Pap smear of CIN 1. Note large, dark nuclei, but also large amount of surrounding cytoplasm.

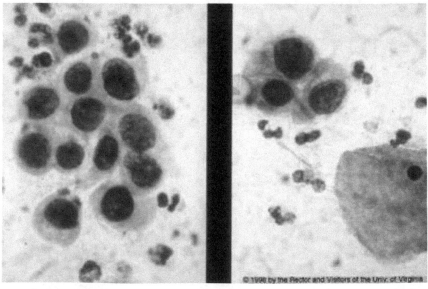

Fig. 13. Pap smear of CIN 3. Note large, dark nuclei with a lesser amount of surrounding cytoplasm. Compare to superficial cell (lower right hand corner).see fig 12 & 13

5. Clinical presentation

5.1 Clinical appearances

CIN lesions are characterized by the appearance of white patches on the cervix following application of acetic acid. Distinct vascular patterns can be seen on colposcopic examination of the cervix in high grade CIN. Lesions occur on the anterior lip twice as commonly as the posterior lip. They are found in the transformation zone and areas of squamous metaplasia in the endocervix and stop abruptly at the junction with the native portio squamous epithelium but can extend along the entire endocervical canal. In general, the portion of CIN on the portio surface is low grade (CIN 1) whereas the portion that extends into the endocervical canal is high grade (CIN 2 and 3).

5.2 Clinical behavior

CIN may regress (spontaneously, especially CIN1), persist or progress. If untreated, up to 16% of CIN1 will progress to CIN3 and up to 70% of CIN3 will progress to invasive squamous cell carcinoma in 1 to 20 years. It is not presently possible to predict which lesions will progress. However, the risk of progression to invasive cancer increases and the time required is shorter with increasing severity of the lesion.(UVa Health).

6. Screening

The aim of screening is to prevent the development of cancer. For screening to be effective, a disease should satisfy the following criteria.

Be common, serious and an important public health concern for the individual and the community.

The disease condition must have a long, latent interval in which pre-malignant change or occult cancer can be detected for the case of cancer of the cervix it is 10-15years.

The natural history of the disease, especially, its evolution from latency to disease should be adequately documented.

There should be effective treatment for pre-malignant change or condition.

The good news is that cervical cancer screening has satisfies the above criteria, especially with regards to developing countries where it really is a public health problem. Cervical screening has been shown to be effective in several countries.

Cervical cancer prevention efforts worldwide have focused on screening women at risk of the disease using Pap smears. Treating precancerous lesions has also prevents cervical cancer in many of the developed countries. In view of the afore mentioned cancer of the cervix is almost extinct in the developed nations,making it the 11th cancer in women and 2nd commonest in developing nations.

6.1 Coverage of the screening programme

It is recommended for all women; especially aged 20 – 64 are invited for screening.

It should be carried out every 3-5 years.

The screening is carried out every 3years in Women aged 45years and bellow.

Where as it is done every 5years in Women aged >45yrs.

Some other risk factors that have been found to be important in developing CIN that would benefit from screening includes⊗(Kumar et al 2007)

- Women who become infected by a "high risk" types of HPV, such as 16, 18, 31, or 45
- Women who have had multiple sexual partners
- Women who smoke
- Women who are immunodeficient and Women who give birth before age 17years.

6.2 Screening technique/process

There are various types of screening test.AS was earlier discussed in this chapter cervical cancer is one of the cancers that has meet all the requirement for cancer screening.The methods that can be employed for this purpose includes,visual inspection using either Acetic acid or Lugos Iodine,Cytological analysis and Human papilloma virous immune assays.

6.2.1 Types of visual inspection test

Visual inspection with Acetic Acid or Visual Inspection with Lugols Iodine.The former is the one that is commonly use for ease of interpretation.

6.2.2 Visual inspection with acetic acid (VIA) can be done with the naked eye (also called cervicoscopy or direct visual inspection [DVI]), or with low magnification (also called gynoscopy, aided VI, or VIAM).

Visual Inspection with Acetic Acid (VIA) – It more relevant in the developing Nations.

Visual inspection with acetic acid (VIA) is an attractive screening method for early-   phase cervical cancer in underdeveloped countries.It is an acceptable screening method for cervical cancer and seems to be an efficient and cost-effective method to detect high-level dysplasia.

6.3 Test performance: Sensitivity and specificity (Defn)

Sensitivity: The proportion of all those with disease that the test correctly identifies as positive.

Specificity: The proportion of all those without disease (normal) that the test correctly identifies as negative.

In the screening of cervical cancer, the sensitivity of VIA was high, whereas the corresponding specificity was only at an acceptable level. The Positive Predictive Value (PPV) and Negative Predictive Value of VIA were found to be high. In other words, the validity of VIA during early-phase screening is high in terms of sensitivity and acceptable for specificity and predictive values. (Ardahan et al).

6.4 Technique of VIA

Performing a vaginal speculum exam is the first step; then the health care provider applies dilute (3-5%) acetic acid (vinegar) to the cervix.

Abnormal tissue temporarily appears white when exposed to vinegar.

The cervix is viewed with the naked eye to identify color changes on the cervix.fig 14 & 15

Determining whether the test result is positive or negative for possible precancerous lesions or cancer and this based on the Aceto-white reactions.

VIA Category	Clinical Findings
Test-negative	No acetowhite lesions or faint acetowhite lesions; polyp, cervicitis, inflammation, Nabothian cysts.
Test-positive	Sharp, distinct, well-defined, dense (opaque/dull or oyster white) acetowhite areas — with or without raised margins touching the squamocolumnar junction (SCJ); leukoplakia and warts.
Suspicious for cancer	Clinically visible ulcerative, cauliflower-like growth or ulcer; oozing and/or bleeding on contact.fig 16

Negative VIA

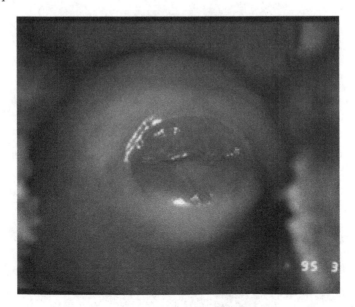

Fig. 14. Photo source: JHPIEGO

Positive VIA

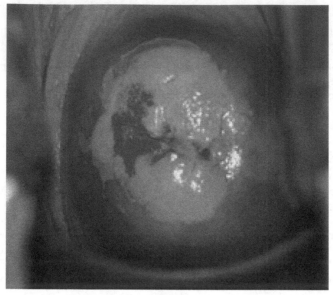

Fig. 15. Photo source: JHPIEGO

Suspicious Cancer

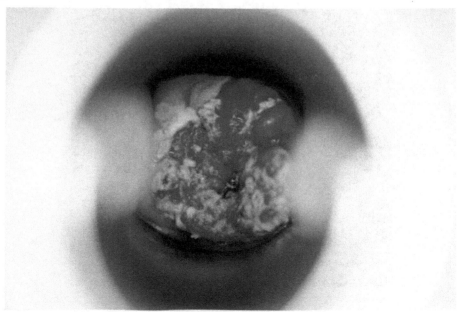

Fig. 16. Suspicion of carcinoma of the cervix. Photo source: PAHO, Jose Jeronimo

6.5 The advantages

It is Simple, easy-to-learn approach that is minimally reliant upon infrastructure.

It is not expensive to start-up and the sustaining costs is affordable.

Many types of health care providers can perform the procedure especially the middle cadre of health care providers.

The Test results are available immediately and as such the issue of follow up is out of the question.

Only requires a single visit.

It may be possible to integrate VIA screening into primary health care services and it will go along way to reduce the incidence and prevalence of carcinoma of the cervix.

There is a need for developing standard training methods and quality assurance measures.

This method isLikely to be less accurate among post-menopausal women caution in its interpretations.

6.6 Visual inspection with Lugol's iodine (VILI)

Visual inspection with Lugol's iodine (VILI), also known as Schiller's test, uses Lugol's iodine instead of acetic acid and it is based on colour change also.

6.7 Pap smear

The Pap test was developed by Dr George Papanicolaou an American anatomist in 1944. Pap test is used primarily as a tool for screening healthy women for preinvasive cervical cancer (CIN) and early invasive cancer.In as much as pap test is a screening tool,it could also be use to identify women at risk of cervical cancer. Women with early invasive cancer (FIGO Stage 1) are often unaware that they are harbouring the tumour as they are usually symptom free. Diagnosis and treatment of invasive cancer while it is still in the early stages of development signficantly improves the prognosis (chances of long term survival) of the patient.

It has been proven over time that the cervical smear may be negative even in the presence of an advanced invasive cervical cancer. This is because blood, inflammatory cells and necrotic debris from the   cancer site frequently obscure the abnormal cells in the smear.

6.8 Specimen sampling

The sample for pap smear can be collected in three ways

6.8.1 a) liquid-based cytology (LBC) - using a cyto- brush a device which samples both endo and ectocervix. These can be used for preparing conventional smear. Some devices have been modified for the preparation of liquid based cytology (LBC) specimens

6.8.2 b) Papanicolaou (Pap) smear test uses a brush or the Ayres spatula to sample the ectocervix. Scraping the ectocervix with with a modified spatula (the Ayre spatula or a variation of it). This is the most widely used method in developing countries and some part of Europe for obtaining material for preparing conventional cervical smears

6.8.3 c) Using an endocervical brush to sample the endocervix this grossly inadequate and it is been discouraged.

Some of the items required for Pap smear.

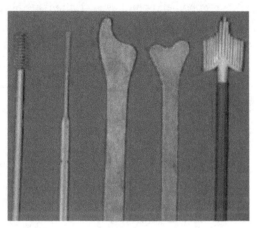

Fig. 17. Example of Fixatives

95% ethanol (for fixation)
80% isopropanol
95% denatured alcohol

Ayres/Cytobrush

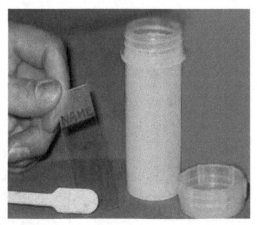

Fig. 18. Fixative Jar/Glass slide

6.8.4 The step by step approach of Pap test

1. A speculum must be inserted into vagina and the cervix clearly visualised.The cervical os should be located.
2. The sampling device(s) used should be selected according to the shape and size of the cervix and the location of the squamocolumnar junction. An Ayre spatula is suitable for sampling the cervix in a parous woman ; however a spatula and brush may be needed in a post menopausal woman where the squamocolumnar junction lies within the endocervical canal.fig 17
3. The pointed end of the spatula should be inserted into the cervical os in a nulliparous cervix and the rounded end of the spatula inserted into the patulous os of a parous woman. The device should be rotated 360 degrees to remove the cells from the region of the transformation zone, squamocolumnar junction and endocervical canal.
4. The material on the spatula or brush must be transferred immediately to a glass slide which has been previously labeled with the patient's name and date of birth.
5. The glass slide (fig 18) must be fixed immediately with an appropriate fixative (95%alcohol) and the slides transported to the cytology laboratory in a container for processing together with the corresponding cytology request form.
6. Samples taken for Liquid Based Cytology should be processed strictly in accordance with the manufacturers instructions. After sampling the cervix, the tip of the sampling device should be broken off into the transport medium in the container provided which should then be transported to the laboratory for processing if the Surepath method is being used. However if the Thinprep method is being used it is of the upmost importance that the tip of the sampling device is not included in the container.

Fixation must be immediate. The smear must not be allowed to dry before fixation.

Test Limitations as it relates to the sensitivity and specificity and technique of smear. There is no difference in specificity, but sensitivity is 12% better with LBC compared with the Pap smear, and its "inadequate rate" is only 1.6%, compared with mean of 9.1% with Pap smears(Sasieni P etal). Problems include:

*   Variable sampling of appropriate cells from the cervix.
*   Poor transfer of cellular material on to the glass slide.
*   Sub-optimal preparation and fixation by the smear taker.

6.9 Smear reporting

It is widely acknowledged that the criteria and terminology used to interpret and report cervical smears differs country to country. This has led to problems of communication between cytopathologists , cytotechnologists and clinicians and makes it difficult for epidemiologists to make valid comparisons of the effectiveness of the different cervical screening programmes. The variability in terminology impedes meaningful discussion between laboratories and also affects patient management and the introduction of optimal methods of patient care.

We have the Following reporting methods:

*   a) British Society: In the UK current reporting guidelines are based on those published by the British Society of Clinical Cytology (BSCC) in 1985
*   These are currently under review and new guidelines are expected soon.

- b) American (Bethesda) system.: The system of terminology used in the United States
- First devised in 1988, being revised in 1991 and again in 2001
- A 2 -tier system which refers to :

Atypical Squamous Cells of undetermined significance **(ASC-US)**

Low grade Squamous intraepithelia neoplasia**(LSIL)**

High grade Squamous intraepithelial neoplasia**(HSIL)**

- Proposed new BSCC guidelines will bring it in line with the Bethesda system

Compare and contrast the two (British & American)

BSCC	BETHESDA
Inadequate	Unsatisfactory for evaluation
Borderline Nuclear Change	ASC-US
	ASC- cannot exclude high grade
	Atypical glandular cells
Mild Dyskaryosis	LSIL
Moderate Dyskaryosis	HSIL
Severe Dyskaryosis	HSIL
Severe Dyskaryosis/?SCC	SCC
? Glandular Neoplasia	Endocervical Ca in situ
	Adenocarcinoma – Endocx., Endomet., Extrauterine, NOS

Fig. 19. Courtesy of Vanessa Jackson

7. Cytology report

7.1 Handling of cytology reports

- Women with normal smears are offered re-screening at the standard 3-5 year interval - follow the advice of your local laboratory. High risk individuals may be screened more frequently. Women with moderate or severe dyskaryotic tests need colposcopy ±biopsy. Women with borderline or mildly dyskaryotic smears are monitored at a reduced screening interval with persistent abnormalities (including persistently inadequate smears) needing colposcopy.

- Unsatisfactory - Repeat smear four weeks later.

7.1.1 Inadequate slide

An inadequate slide may occur as a result of insufficient smear, or when it is adequate it is obscured .As much as possible this should be avoided.

In phase of any abnormal cells, a result of inadequate should not be given rather the abnormality should be reported.

7.1.2 Management of inadequate smear

It is an indication for a Repeat smear

- If there is a recognisable infection present the patient should be treatmented before a repeat smear is done.
- After three consecutive inadequate samples, their is a need to refer such for colposcopic assessment.

7.1.3 Negative

There should be enough cellular material to cover 1/3 of the slide before a pronouncement or report of a negative smear. The report actually makes the woman to be confident that she is not at risk of any dysplasia for a period of three years.

- In atrophic smears, where the cellular material is comprised of parabasal sheets, 10% of the slide can be considered adequate.
- There are no official guidelines for LBC samples.

It is possible that 15,000 cells will be the standard for adequacy.

7.1.4 Management of negative slides

We should give a recommendation of routine recall to our clients. They should be encouraged to adhere to this recommendation:

- Women Age 25-45 should be advised to repeat there smear *every* three *years.* While those above 45years to 65 should have smear every five years.

Exceptions: There are exceptions to the above stated rules or recommendation.

Any patient with Clinical symptoms which are suggestive of immune surpression or are immune compromise or have been diagnosed to have immune compromised disease such as HIV positive women or women on immune suppressive drugs as in patients with renal transplant should have a repeat smear every year/12 *months*

The other exception is for those women who may have had the following conditions in the past : Post coital bleeding (PCB),Post menopausal bleeding(PMB),and friable cervix

All patients on follow up for previous abnormal smears are candidate for repeat based on the findings.

7.1.5 Reporting of infections

The presence of specific infections may be reported based on the histological features of such infections: Such as

- Candida
- Trichomonas – with advice to culture before treatment
- Herpes Simplex Virus – referral for counselling should be advised
- Actinomyces – like organisms
- Follicular Cervicitis – report in younger women

It is advisable that additional investigations towards definitive diagnosis must be pursued before embarking on treatment.

7.1.6 Follow up of patients on treatment

The Follow-up of women who have been treated for CIN is very crucial to ensure that there is no progression of the disease condition.

- CIN1 – repeat smears at 6months, 12months and 2years this is the schedule if the smear is persistently negative.
- CIN2 and above –This categories of patients require annual smears for 10 years of follow up.
- CGIN(Cervical Glandular Intraepithelia Neoplasia) – are at greater risk of recurrent disease so they are recommended to have smears every 6months for 2years,then annually for 10 years

Follow-up of women with low grade smears but normal colposcopy and no biopsy – require a repeat smear at 6 months, 6 months , 12 months, then return to normal recall

Follow-up of women after hysterectomy for CIN or SCC :

Where there was complete excision and the margins were clear of dysplastic cells, vault smears should be carried out at 6 months and 18 months before recall can be cancelled as no further smears are required.

In the case of women with incomplete or uncertain excision at hysterectomy, they would require follow-up as for women with CIN2 or above.

In women who have been exposed to radiotherapy as an adjuvant therapy, Smears are not advised in this group of women.

7.1.7 Borderline nuclear change

This is a "holding" category where there is genuine doubt as to whether or not a smear is abnormal. The nuclear changes are not typical and convincing of cervical dyskaryosis and as such it can not be labelled as negative.

Borderline nuclear change is used when wart virus changes are seen on the smear, without dyskaryosis.

It can also be used when severe inflammatory changes exist on the smear, which can sometimes appear almost dyskaryotic.

When interpretation of the smear is difficult e.g. as it is in poor handling or fixing and or staining process (such as due to air drying)

7.1.8 Management of borderline nuclear change

The smear should be repeated at 6 months, interval for 1year and 12 months later. – If all are negative, normal recall can be resumed.

If in the course of the follow up, there are a maximum of 3 reports of borderline nuclear change in the follow-up period, referral for colposcopy is advised.

At any point in time One report of borderline glandular cells requires immediate referral for further evaluation..

In difficult cases, where there is concern that high grade disease may be present, immediate referral can be recommended.

8.1 CIN I/Low sil management

CIN I/Low SIL correlates to Nucleus occupying up to 1/2 of the area of the cell(Nucleocytoplasmic ratio of half)

8.1.1 Management of mild dyskaryosis

It is advisable that she should have a colposcopy done,in centres where this facilities are not available, "it remains acceptable to recommend a repeat test".

If the repeat smear is the same diagnosis (mild dyskaryosis) then a referral must be advised.

8.2.1 CINII/High SIL

8.2.2 Cytological features

CIN II/High SIL correlates to Nucleus occupying up to ½ to 2/3 of the area of the cell(Nucleocytoplasmic ratio of ½ to 2/3)

The Chromatin pattern is usually more abnormal compare to CIN I

If a smear is obviously dyskaryotic, but difficult to grade e.g. because there are too few cells or they are poorly preserved, it is recommended that they be coded as moderate.

8.2.3 Management of moderate dyskaryosis

All patients with moderate dyskaryosis should be referred for colposcopy.

8.3.1 CIN III/ High SIL

8.3.2 Cytological features

CIN III/High SIL correlates to Nucleus occupying more than 2/3 of the area of the cell(Nucleocytoplasmic ratio greater than 2/3).The nucleus may have a bizarre shape.

The Chromatin pattern is usually more abnormal compare to CIN I

8.3.3 Management of severe dyskaryosis/CIN III/High SIL

Referral for colposcopy is the standard approach of management.This will include tissue biopsy for histology.

8.4.1 Invasive squamous carcinoma

8.4.2 Cytological features

The histological features are essentially that of Bizarre nuclear changes and keratinisation.

Diathesis is also a notable findings in this patients.

8.4.3 Management of invasive squamous carcinoma

In this case an URGENT referral for colposcopy and tissue diagnosis is advised.

9. Treatment of cervical intra-epithelia neoplasia

9.1 Low grade lesions

In the treatment of this disease entity a colposcopy,with or without a repeat smear,and or tissue biopsy is an essential requirement as stated above. The uses of additional investigative tools are very essential in the treatment of this condition. The time interval between diagnosis and treatment can be very crucial

The option of treatment range between Cryotherapy, cold coagulation,Laser agglutination therapy and Electrocautery. A lot of caution must be applied to avert over treatment especially in young women who are still desirous of conception (over treatment can cause fertility problems).

The follow up schedule as stated above and the patient should be encouraged to adhere to this to achieve the desired goal of screening.

9.2 High grade lesions

The additional investigations include Colposcopy, repeat smear and tissue biopsy is important toward establishing a diagnosis,because the treatment involved is usually irreversible. Such   definitive treatment   includes ablative procedures and amputation surgeries.

The definitive treatments include Cold coagulation,Lletz, laser agglutination therapy, electrcautery, knife cone biopsy and trachelectomy.

9.3 Summary of indication for colposcopy examinations

The under listed are the patient that would benefit from colposcopy

- Following 3 consecutive inadequate smears.
- Women with clinical symptoms e.g. PMB, friable cervix
- Post menopausal women with unexplained endometrial cells
- Women with genital warts.
- Persistent borderline smears (maximum 3)
- One report of borderline nuclear change in glandular cells.
- After 3 abnormal smear results (any grade) in a 10 year period.
- One report of mild dyskaryosis.

- One report of moderate dyskaryosis.
- One report of severe dyskaryosis.
- One report of ? invasive SCC.
- One report of ? glandular neoplasia.
- A report of mild or worse in women on follow-up for treated CIN

10. Human immunosuppressive virus and cervical intra-epithelia neoplasia

This is actually of more relevance in the developing nations. It's a major scourge of our time that needs to be address at all time and at every given opportunity.

Follow up in CIN cases is done closely in HIV-positive women: treatment of CIN I has a high failure rate in these women, but it has a relatively low rate of progression (Robert Finn 2011).

HIV-infected women were 3.7 times more likely to develop CIN than HIV-uninfected women. These results highlight the importance of regular cervical cytological screening for HIV-Infected women.(Wright)

The interplay between HIV infection, HPV infection and CIN/cervical cancer is complex (see Box 34.1). Cervical dysplasia and possibly invasive cervical cancer are more prevalent in HIV-positive women. The latter has a higher rate of HPV infections which are strongly associated with high-grade SIL and invasive cervical cancer.(Sun et al 1997) Immune suppression is the factor predisposing HIV positive women to HPV infections fig 20. CIN is commoner in HIV infected women with a lower CD4 count or AIDS. (Sun et al 1995) have suggested that the presence of immunosuppression shifts the ratio of latent: clinically expressed HPV infection from 8:1 in the general population to 3:1 in HIV-positive women with CD4 >500/µL and to 1:1 when CD4<200/µL. Linking the US AIDS and Cancer Registry, the observed cervical cancer cases in HIV infected women were up to 9 folds higher than the expected number of cases but the likelihood of cervical cancer is not related to the CD4 count. (mbulaiteye et al2003).

The screening process for the HIV positive patients differs significantly based on the prelude of our discussion and as such the have the following schedules:

Pelvic examination and Pap smear are repeated six months after the baseline.(Anderson 2005) If both times are normal, cervical screening is then performed every twelve months alongside with careful vulval, vaginal and anal inspections. There is no specific CD4 threshold under which the frequency of Pap smear would need to be increased, though this may be considered in cases of(Anderson 2005)

a. Previous abnormal Pap smear
b. HPV infection
c. Symptomatic HIV disease
d. CD4 counts <200/µL
e. Post-treatment for CIN

Other forms of screening may be employed in the management of those who are positive for HIV. *HPV-DNA* has recently been introduced as one form of screening. As there is a high incidence of HPV infection in HIV patient and most HIV patients with abnormal smear will ultimately need colposcopy evaluation (to rule out CIN and cervical cancer), the use of HPV-DNA in the triage for colposcopy is of limited value and it is not cost effective. On the other hand, despite the higher incidence of cervical abnormality in HIV infected patients, *colposcopy* is not generally recommended for primary cervical screening (Anderson et al 2005) but indicated when pap smear reveals epithelial cell abnormality.

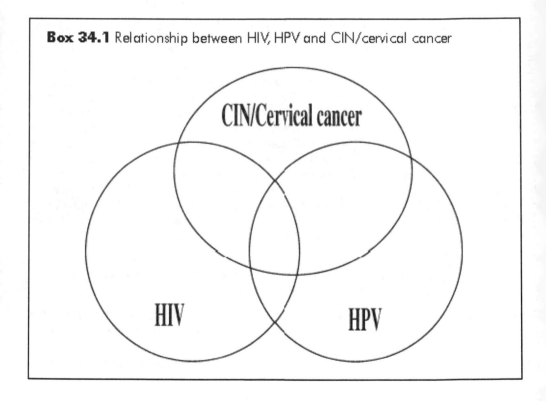

Box 34.1 Relationship between HIV, HPV and CIN/cervical cancer

CIN/Cervical cancer

HIV

HPV

Fig. 20. (Courtesy Siu-Keung LAM) Relationship Between HIV, HPV & CIN 1

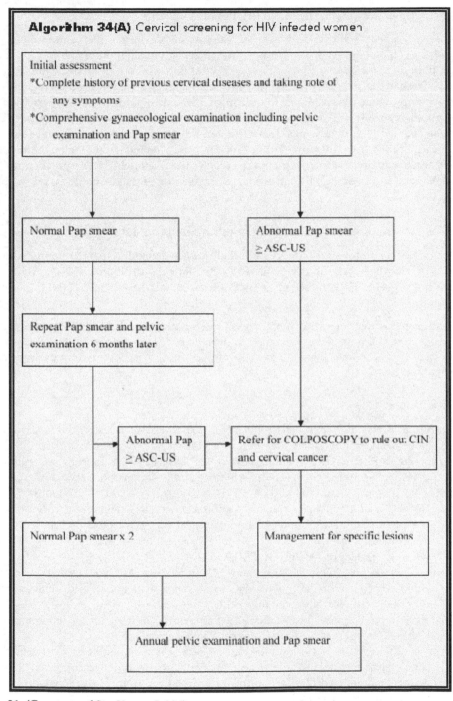

Fig. 21. (Courtesy of Siu-Keung LAM)

10.1 Management of cervical lesion in HIV positive patients

HIV-infected patients with ASC-US cervical lesion or above once on Pap smear should be referred for colposcopy. At a colposcopy clinic, thorough evaluation of the lower genital tract is performed including colposcopy and cervical biopsy at the most suspicious area to rule out CIN and/or malignancy. If there is no HGSIL, patients are followed up regularly in the colposcopy clinic. If HGSIL is found, most patients require large loop excision of transformation zone (LLETZ), to reduce the chance of progression to cervical cancer. This has to be coupled with regular post-treatment cervical smear surveillance as stated above. The other forms of therapy that can be use includes, Laser agglutination therapy,Electrocautery and cone biopsy of the cervix. The role of HAART in preventing CIN recurrence post treatment is still controversial,but it is adviced that it should be given.(**Siu-Keung LAM**)

11. HPV vaccination and cervical intra-epithelia neoplasia

In a randomised control study (double blinded) it was concluded that, In young women who have not been previously infected with human papillomavirus-16 (HPV16), vaccination prevents HPV16-related cervical intra-epithelial neoplasia (CIN).(Mao et al 2006).

It should be noted that only 75% of all cervical cancers are caused the HPV viruses 16 and 18,it is therefore still possible for a woman to develop cervical cancer even though they are immunised. This is because there are other sero types of HPV not covered by those vaccine in the market.

12. References

ACCP. Visual screening approaches: (October 2002). Promising alternative screening strategies. Cervical Cancer Prevention Fact Sheet.

ACCP & World Health Organization. November 2003 Cervical cancer prevention in developing countries: A review of screening and programmatic strategies.

Agorastos T, Miliaras D, Lambropoulos A, Chrisafi S, Kotsis A, Manthos A, Bontis J (2005). "Detection and typing of human papillomavirus DNA in uterine cervices with coexistent grade I and grade III intraepithelial neoplasia: biologic progression or independent lesions?". *Eur J Obstet Gynecol Reprod Biol* 121 (1): 99–103. doi:10.1016/j.ejogrb.2004.11.024. PMID 15949888.

American College of Obstetricians and Gynecologists.(ACOG) Obstet Gynecol. 2010;Cervical cancer in adolescents: screening, evaluation, and management. Committee Opinion No. 463. 116:469–72.

America Society for colposcopy and Cervical Pathology (ASCCP) 2011 Anatomy of the Uterine Cervix.The society for the lower Genital Tract disease.

Anderson JR. 2005 edition. A guide to the clinical care of women with HIV-. Published by US Department of Health and Human Services, Health Resources and Service Administration, HIV/AIDS Bureau. Available from http://hab.hrsa.gov/publications/womencare05.

Ardahan, Melek PhD, RN; Temel, Ayla Bayik PhD, RN. March/April 2011 Visual Inspection With Acetic Acid in Cervical Cancer Screening Cancer Nursing: Volume 34 - Issue 2 - pp 158-163doi: 10.1097/NCC.0b013e3181efe69f Articles.

Barbara G,2011 Cervical Intraepithelia Neoplasia in Up to Date Marketing Professionals.

Dina.R. Eurocytology 5th-10th September 2011 Hammersmith AdvancedClinical Cytopathology Course Held at Imperial College- HammersmithHospital campus - Commonwealth Building.

IARC(International Agency for Research on Cancer) 2001 An Introduction to Cervical Intraepithelial Neoplasia (CIN)charpter 2.

Kumar, V; Abbas, A K.; Fausto, N; & Mitchell, R N. (2007). Robbins Basic Pathology (8th ed.). Saunders Elsevier. pp. 718–721. ISBN 978-1-4160-2973-1. ^Cervical Dysplasia: Overview, Risk Factors.

Mao C, Koutsky LA, Ault KA, et al. 2006 Efficacy of human papillomavirus-16 vaccine to prevent cervical intraepithelial neoplasia. Obstet Gynecol; 107:18-27.

Mark H (2010)UNSW Embryology of Genital System - Female Uterus

Mbulaiteye SM, Biggar RJ, Goedert JJ, Engels EA. 2003 Immune deficiency and risk for malignancy among persons with AIDS. J Acquir Immune Defic Syndr;32:527-33.

NEL J. T.; DE LANGE L. ; MEIRING P. J. ; DE WET J. I.1994 Cervical intra-epithelial neoplasia and invasive cervical cancer in black and white patients ; SAMJ. South African medical journal   ISSN 0256-9574   CODEN SAMJAF , vol. 84, no1, pp. 18-19 (16 ref.)

Oguntayo O A and Samaila .M O A Prevalence of Cervical Intraepithelia Neoplasia in Zaria,Annals of African Medicine Vol 9,No 3;2010:194-5.

Robert Finn 2011Follow CIN closely in HIV-positive women: .(Gynecology): An article from: OB GYN News [HTML] [Digital.

Sasieni P, Adams J (1999 May):Effect of screening on cervical cancer mortality in England and Wales: analysis of trends with an age period cohort model;*BMJ* 8;318(7193):1244-5.

Siu-Keung LAM CERVICAL NEOPLASIA IN HIV/AIDS.

Sun XW, Ellerbrock TV, Lungu O, Chiasson MA, Bush TJ, Wright TC Jr. 1995 Human papillomavirus infection in human immunodeficiency virus-seropositive women. Obstet Gynecol;85(5 Pt 1):680-6.

Sun XW, Kuhn L, Ellerbrock TV, Chiasson MA, Bush TJ, Wright TC Jr. 1997 Human papillomavirus infection in women infected with the human immunodeficiency virus. N Engl J Med;337:1343-9.

UVa Health 2006 Pathology Thread IV. *Premalignant and Malignant Neoplasms* University of Virginia School of Medicine. Department of Patholo PO Box 800214 System www.med-ed.virginia.edu/courses/path/gyn/cervix4.cfm -

Vanessa Jackson 2006 CERVICAL CYTOLOGY REPORTING AND TERMINOLOGY AND MANAGEMENT OF ABNORMAL SMEARS Advanced Biomedical Scientist Practitioner Leeds Teaching Hospitals NHS Trust.

Wright TC, Ellerbrock TV, Chiasson MA, Williamson J, Sun X, Bush TJ 1995 Jan 29-Feb 2 Incidence and risk factors for cervical intraepithelial neoplasia (CIN) in HIV-infected women.; National Conference on Human Retroviruses and Related Infections. *Program Abstr Second Natl Conf Hum Retrovir Relat Infect Natl Conf Hum Retrovir Relat Infect 2nd Wash DC*; 90. Columbia University and NYC Department of Health, NY.

Wart From Wikipedia, the free encyclopedia www.facebook.com/pages/Warts/110511208977959

Wright TC Jr, Cox JT, Massad LS, Twiggs LB, Wilkinson EJ; 2002 ASCCP-Sponsored Consensus Conference. 2001 Consensus Guidelines for the management of women with cervical cytological abnormalities. JAMA;287:2120-9.

4

P16INK4A and MIB-1 Expression in Preneoplasia and Neoplasia of Cervix

Supriya Srivastava
Cancer Science Institute, National University Singapore,
Singapore

1. Introduction

Cervical cancer is the third most common cancer and fourth most common cause of cancer related deaths in female population, accounting to approximately 453,300 cases per year and 275,100 deaths in the year 2008. According to the latest WHO global cancer statistics, the cumulative risk (%) (Age 0-74) of cervical carcinoma is 0.9 with age adjusted ratio of 9.0 (Jemal et al., 2011). In India, cervical cancer is the leading cancer among females between 15 and 44 years of age. Current estimates indicate that every year 132,082 women are diagnosed with cervical cancer and that 74,118 die from this disease in India alone (http://www.who.int/hpv). Having said this, however, no form of cancer better documents the remarkable effects of prevention, early diagnosis, and curative therapy on the mortality rate than the cancer of cervix. However, the still very high rate of cervical carcinoma in developing countries like India is because of lack of proper screening methods and lack of health infrastructure which allows for periodical and routine screening. Potential threat of cancer has reduced significantly in developed countries, due to Papanicolaou smear screening programs. Papanicolaou smears or commonly referred as Pap smears is a cost effective and reproducible screening technique for diagnosing precursor lesions of cervical carcinoma. However, Pap test gives significant false positive (30%) (Sherman et al., 1994) and false negative (15-50%) results due to subjective test criteria (Arbyn et al., 2008). Apart from the Pap smear screening test, histopathological diagnosis of cervical intraepithelial neoplasias (CINs) and cervical carcinoma is considered as the age old "gold standard" method of diagnosis of cervical neoplasms. However this can also be biased by interobserver variability as reported before (Stoler & Schiffman, 2001). These factors limit, present screening programs and histopathological examination and emphasizes the need for the identification of specific biomarkers for dysplastic epithelial cells to aid in primary screening and lesion diagnosis.

2. Cervical Intraepithelial Neoplasia (CIN)

Invasive squamous cell carcinoma of cervix is preceded by precancerous changes in the cervical epithelium which can be identified histologically. These precancerous lesions are usually described as Cervical Intraepithelial Neoplasia (Buckley et al., 1982). Papanicolaou classification, using the terms 'atypical cells with abnormal features' has been adhered to, until recently by some cytologists and gynaecologists. In 1953 Reagan *et al.* proposed the term **"dysplasia"** to replace atypical metaplasia and atypical hyperplasia. Ritton and Christopherson defined the normal and abnormal cells of cervical and vaginal smears in the WHO International classification (1973). The conventional histological terminology of mild, moderate and severe dysplasia and carcinoma in situ was used as well as atypical metaplasia. The British Society for clinical cytology's first Working party on terminology recommended the term **"dyskaryosis"**, originally coined by Papanicolaou and translated from the Greek meaning "abnormal nucleus", to describe cells from preinvasive and invasive cancer (Spriggs et al., 1978). In 1986, in a further review, dyskaryosis remained the recommended term, but it was classified as *mild, moderate* and *severe*.

The 1988 Bethesda System for reporting Cervical/Vaginal Cytologic Diagnoses was published by a Workshop of North American Experts convened by the Division of Cancer Prevention and Control of the National Cancer Institute to review existing terminology and to recommend effective methods of reporting. It agreed that the Papanicolaou classification was no longer appropriate and proposed the **Bethesda System,** which recommends three essential components of a cervical or vaginal smear report. It includes a new term, **Squamous intraepithelial neoplasia** (SIL) which is divided into two grades, **low grade SIL** (cells from HPV and CIN-I) and **high grade SIL** (cells from CIN II and CIN III) (Broder et al., 1991). A Bethesda workshop was held in 2001 with further modifications. The descriptions of cytological appearances of the cells of precancerous conditions of the cervix are best understood in relation to the well defined three histological grades of CIN (Solomon et al., 2002).

2.1 Low grade Squamous Intraepithelial Lesion (LSIL)

In LSIL the cells are mature squamous cells, they retain their polygonal shape and for the most part retain their normal size with a peripheral rim of dense cytoplasm. The nuclei are enlarged at least 3-4 times that of the normal intermediate cell nucleus, however, when HPV changes are evident, the cells may be smaller (almost parakeratotic) and the nuclei may also be smaller and somewhat pyknotic appearing with binucleation and/or multinucleation. These pyknotic nuclei also exhibit abnormal features such as hyperchromasia, increased size from that of the normal superficial squamous cell and a slight variation in shape and size. The chromatin appears finely to coarsely granular and is evenly distributed. It is important to stress that an interpretation of LSIL/HPV requires both clear-cut cytoplasmic cavitations accompanied by the abnormal nuclear morphology described above.

Differential diagnoses

Reactive

The cells appear single or in sheets, like LSIL, however unlike LSIL, where only mature squamous cells are affected, in reactive types, all the cells may be affected. The nuclei, may be enlarged from 1.5 to 2 times; bi or multinucleation may be present and the nuclear membrane appear smooth. The chromatin is finely granular, evenly distributed and hyperchromasia may not be evident. The nucleoli are uniform and may be multiple in numbers. Peri-nuclear halo is often present, small and multiple vacuoles may be evident due to degeneration

Reparative changes

The cells appear in flat sheets or groups. Predominantly, endocervical or metaplastic cells are affected. The nuclear size may be variable, ranging from slight to marked enlargement. The nuclei may be bi or multinucleated, and the nuclear membranes appear smooth. The chromatin is finely granular and evenly distributed and hyperchromasia may not be evident. The nucleoli are small to conspicuous and often multiple. The cytoplasm appears vacuolated.

2.2 High grade Intraepithelial Lesion (HSIL)

The criteria for HSIL on the ThinPrep® Pap Test are as follows: The single, most important criterion for HSIL is the presence of asymmetrical 3-D nuclear structural abnormalities. This is a concept that must be clearly understood in order to master the interpretation of HSIL. There will be an abnormality in the structure of the dysplastic nucleus that can be thoroughly appreciated only by focusing up and down on the individual cell. A normal nucleus has a relatively round or ovoid shape and its surface is smooth. A dysplastic cell will have humps, bumps, corrugations, crevices, and strange protuberances. These very distinctive abnormalities are the essence of dysplasia, particularly HSIL. This is the very detail that is most often lost in conventional cytology due to the various artefacts of fixation and staining that limit the ability to interpret these conventional smears. These 3-D nuclear structural abnormalities are to be distinguished from simple, "irregular nuclear outlines" which will often be present as a two-dimensional phenomenon in benign cells on the ThinPrep® Pap Tests.

These 3-D structural abnormalities may not be present in every dysplastic cell on the slide, but they will be obvious in at least some cells somewhere on the slides. Obviously, the ability to see "into" the nucleus of a cell is going to be directly related to the quality of staining. Also, these 3-D structural defects should be asymmetrical, as opposed to nuclear grooves or simple creases that involve the full breadth of the nucleus occasionally creating a difficult "look-alike". The presence of these exaggerated nuclear 3-D abnormalities establishes the diagnosis of HSIL.

Apart from above the N/C ratio is the most reliable indicator of degree (moderate, severe, CIS). With an increasing degree of dysplasia, there is a predictable increasing N/C ratio. This abnormal N/C ratio can be considered a major criterion for the diagnosis of HSIL. However, there are rare exceptions, and ultimately the diagnosis must be made on nuclear changes alone.

Gland neck involvement in HSIL has a distinct appearance on the ThinPrep Pap Test and can be differentiated from lesions of glandular origin. SIL in glands presents predominantly in sheets with increased depth of focus. The cytoplasm is finely vacuolated which initially may give the impression of a glandular process, but on closer inspection these sheets exhibit no glandular differentiation such as basal nuclei, crowded columnar formations, pseudostratification, nor feathered edges or rosettes. These sheets of cells can be deceptively flat, but the nuclei retain the same qualities of SIL that are described above. Because these cells are in sheets and usually are small with no other "clues" of HSIL - only subtle 3-D deformities, they can be the most difficult to identify and evaluate. An important factor in determining whether or not these cells are squamous or glandular in origin is the company they keep. Most of the time these cells will be accompanied by definitely dysplastic squamous epithelial cells.

2.3 Atypical Squamous Cells of Undetermined Significance (ASCUS)

ASCUS in the reproductive woman is defined by a number of criteria. The principal one is nuclear size using either an intermediate squamous cell nucleus or a mature metaplastic squamous cell nucleus as the reference standard. An ASCUS nucleus is 2.5 to 3 times the size of an intermediate cell nucleus or 1.5 times the size of a mature metaplastic cell nucleus. Hyperchromasia is commonly present, nucleoli are not prominent. These nuclear features are most important in diagnosing ASCUS.

3. Etiology of CIN and cervical carcinoma

There are various etiological factors leading to CIN and eventually to cervical carcinoma. Active and passive smoking (Brinton et al., 1986), dietary deficiencies (Butterworth et al., 1992), immunosupression (Zur-Hausen, 1993) and sexually associated factors are a few to name. Among these, sexually associated factors (SAF) are the most important in pathogenesis of cervical carcinoma and CIN. Multiple sexual partners, early marriage (Munoz & Bosch, 1989), male sexual behaviour (Brinton et al., 1989), concurrent penile cancer in males (Li et al., 1982) and sexually transmitted diseases are strongly associated with the development of carcinoma. Viral infections with Herpes simplex virus- 2 (HSV-2) (Fenoglio et al., 1982) and Human Papilloma Virus (HPV) has been implicated and studied extensively in the etiopathogenesis of cervical carcinoma and CIN (Meisels et al., 1981)

3.1 Role of HPV in etiology of cervical neoplasia

Innumerable experimental studies have provided strong evidence that HPV is the long sought venereal cause of cervical neoplasia. These viruses are   double stranded DNA viruses (Baltimore Class I) and have been included traditionally in the Papovaviridae (Tomita et al., 1987). HPV is a double stranded Papilloma virus; 70 different types of which are known. Many different HPV types associated with cervical neoplasia have been discovered and around twenty types of HPV are commonly known to infect the human genital tract. These are HPV 6, 11, 16, 18, 30, 31, 33, 34, 35, 39, 40, 42, 51-58 (Crum et al., 1991)· However they have been divided into high- and low-risk categories based on their association with invasive cervical carcinoma (Lorincz et al., 1992). Of this HPV 16, 18 and 31 are more commonly implicated in cervical carcinoma (Zur- Hausen, 1991). Experimental data indicate that viral E6 and E7 genes of high-risk HPV E7 protein specifically bind to and inactivate pRB (retinoblastoma gene product).

3.1.1 Structure of HPV

HPV-DNA consists of three different regions: early region (ER), late region (LR) and upstream regulatory regions (URR). The ER is composed of seven genes, E1-E8, that encodes proteins which play a significant role in viral replication and have oncogenic properties. The LR is composed of two genes L1 and L2, which encodes proteins required for assembly of infectious viral particles. The URR is the regulatory region. In preneoplastic lesion like CINs the HPV DNA is not integrated in host DNA, rather it is found in circular or episomal form. Briefly, the episomal HPV produces mostly the E2 protein. The E2 protein encodes for a DNA-binding protein that binds to a specific nucleotide motif found in E6 and E7 region (Ham et al., 1991) (Ward et al., 1989). There therefore the E2 regulates the expression of E6 and E7, so that only low level of these proteins is produced. The episomal form integrates into the host's chromosome at E1/E2 region, typically causing "break" in this region, giving rise to uncontrolled production and expression of E6 and E7 proteins and their high levels are produced. Scheffner et al have shown previously that E6 protein forms a complex with p53 tumor-suppressor gene product, leading to the degradation of p53. Crook et al studied the expression of p53 in several HPV-positive or HPV-negative cell lines, and found that the HPV-negative cell lines had a mutation in a single nucleotide (a point mutation) in the p53 mRNA. These findings suggest that the loss of the wild type p53 protein activity is important in the development of a malignant lesion, and this could be mediated either by point mutation or by the binding of HPV E6 protein to p53. The HPV E7 protein binds to retinoblastoma tumor-suppressor gene product and inactivates the p Rb (Scheffner et al., 1992). The degradation of p53 and functional inactivation of p Rb leads to cell cycle disruption and increased proliferation, ultimately giving rise to carcinoma.

3.1.2 HPV detection techniques

Detection of HPV DNA in CIN and cervical carcinoma is the most popular and well investigated biomarker in management of cervical neoplasia. Various techniques are used to detect DNA. These are:

1. Immunocytochemistry
2. Dot Blot assays
3. Southern Blot
4. In situ Hybridization
5. Hybrid Capture™ 2 (HC2) assay
6. Polymerase chain reaction (PCR) techniques
7. HPV genotyping
8. Immunocytochemical detection of L1 capsid protein.

Among the above mentioned In situ hybridization. HC2 assays and PCR are most commonly used methods to detect HPV. Currently according to the updated guidelines published by American Society for Colposcopy and Cervical Pathology (ASCCP), the "HPV testing" refers only to Hybrid Capture 2 test for high-risk types (HPV16 and 18) (Wright et al., 2007). The HC2 test is the only test presently approved by U. S. Food and Drug Administration.

4. Biomarkers in cervical neoplasia

Cytomorphological interpretation of Pap stained cervical smears is the mainstay of cytological evaluation of the human cervix. A wide array of potential biomarkers is being evaluated for the diagnostic usefulness of cervical cancer and its precursors. One of the needs to identify biomarkers in cervical neoplasia is to distinguish CIN from other non neoplastic cervical lesions, so as to prevent under treatment (Al Nafussi et al., 1990) or overtreatment (Creagh et al., 1995). Second purpose is, that since CIN is a dynamic process (not a static process), that can progress or regress, the conventional haematoxylin and eosin (H&E), gives a false impression of a static process. These points emphasize the need to identify and discover new markers that can aid in distinguishing CIN from other benign conditions and establish it as a dynamic process. Since HPV, disrupts the normal cell cycle, leading to cell death, a number of genes/ proteins are deregulated, thereby such genes/proteins can be used as surrogate diagnostic markers. In the past few years number of genes/proteins has been implicated as suitable biomarkers for cervical neoplasia. Two markers that have shown a potential in this direction are p16 INK4A and MIB1. p16 is a tumor suppressor protein, that is expressed in dysplastic cervical epithelial cells only, while MIB-1 is a marker of active dividing cells (basal and parabasal cells), normally not shed in cervical smears. Therefore presence of p16 and MIB-1 positivity in cervical Pap smear is marker of cervical dyskaryosis. Due to these reasons, p16 and MIB-1 have emerged as the most robust, stable and markers with strong predictive value.

4.1 P16INK4A

P16INK4A (inhibitor of kinase 4A), is a tumor suppressor protein and inhibitor of cyclin-dependant kinase 4 and 6 (CDK 4 and 6). The phosphorylation of pRB (retinoblastoma protein) is a molecular "ON–OFF" switch for the cell cycle. In the hypophosphorylated form, pRB binds to transcription factors (p 16) responsible for cell cycle progression. P16 inhibits the cyclin-dependant kinases and thereby prevents the phosphorylation of RB, keeping it in the hypophosphorylated form, i.e. its active form. However, in HPV infection, the viral gene E7 binds to RB protein and functionally inactivates it. This results in accumulation of p16 protein because, normally, RB inhibits the transcription of p16 (Keating, 2001; Klaes, 2001; Sano, 1998). Because this protein is not expressed in the normal cervical epithelium, p16 overexpression allows to specifically identify dysplastic lesions and will reduce interobserver disagreement of conventional histological or cytological tests.

4.1.1 P16INK4A as a diagnostic biomarker in cervical neoplasia

p16 INK4a is a tumor suppressor protein (cyclin dependant kinase inhibitor) which is known to play a critical role as a negative regulator of cell cycle progression and differentiation by controlling the activity of tumor suppressor protein pRb. We performed a study on the role of p16 and MIB-1 in cervical intraepithelial neoplasia. Our hypothesis was that normal cervical epithelium does not express p16 INK 4a and MIB-1 and there is upregulation of these biomarkers in CINs and cervical carcinomas. We evaluated p16 and MIB-1 in 63 cervical biopsies and corresponding Pap smears. p16INK 4A immunostaining was done using Mouse monoclonal antibody RTU-p16-432 (Novocastra, Lab. Ltd., Newcastle, Tyne, NE 128EW, UK, in all study groups. Immunopositive was considered when there was Either diffuse, strong nuclear and cytoplasmic staining, or focal moderate to weak nuclear staining of tumor cells. p16 INK4A immunohistochemistry revealed that there was a significant over expression and upregulation in different groups and as we move from normal cervical epithelia to dysplasia of varying severity to carcinoma, the p16 positivity was increased. p16 INK4A over expression was seen in all CIN I lesions (15/15), all CIN II lesions (15/15), all CIN III lesions (3/3) and all cases of carcinoma cervix (15/15) of tissue biopsies. In Pap smears p16 positivity was seen in CIN I/LSIL (8/10), CIN II/HSIL (5/5), CIN III/HSIL (3/3) and Ca cervix (15/15). No detectable p16 expression was observed in normal cervical epithelium in both pap smears and tissue biopsies. This was found to be statistically significant finding on making a comparison between control versus different groups (p<0.05). However, on making an inter group comparison this was found to be statistically insignificant (p>0.05). p16 basically is a nuclear protein hence immunohistochemistry should show nuclear staining. However in dysplasia both nuclear and cytoplasmic staining with p16 is observed possibly because of post transcriptional modification or overproduction of p16 protein forcing its transfer into the cytoplasm (Murphy *et al.*, 2002). In our study it was seen that p16 over-expression was restricted to CIN I, II, III and carcinoma cervix and increased in the same order. Therefore, p16 immunostaining allowed precise identification of even small CIN or cervical cancer lesions in biopsy sections and Pap smears and helped to reduce interobserver variation and also reduce false positive and false negative interpretation and thereby significantly improve cervical cancer and precancer detection (Srivastava, 2010). In our study p16 positivity was 15 of 15 (100%) in invasive carcinoma cervix and it was seen that with increasing severity of CINs, p16 positivity increased. Similar results were seen in a study by Murphy *et al.* who observed 100% p16 positivity in invasive SCC and significant linear relation (p<0.0001) between p16 staining and increasing grades of squamous dysplasia (Murphy *et al.*, 2002, 2005). We also observed that two Pap smears with LSIL showed negative p16 staining whereas it was positive in corresponding CIN lesion of their tissue section. p16 may be rarely negative in cervical dyskaryosis that may have important implications for the use of p16 staining as a standalone test and support the use of combination of markers of cervical dyskaryosis (Murphy *et al.* 2005). However in our study we did not find any dysplasia negative for p16 in tissues biopsies. p16 staining in LSIL was found to be negative in 20% of Pap smears, which could possibly be due to the technical error as their corresponding sections showed consistent positivity.

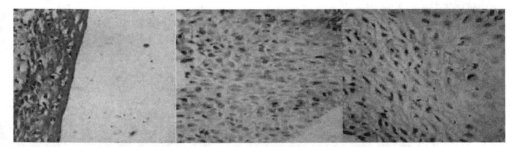

Fig. 1. Images of p16 immunostaining in Cervix tissue biopsy in CINI, CINIII, and Carcinoma cervix (40X)

A study conducted by Trigler *et al.*, in 2004, confirmed that the proportion of pRb positive cells was relatively decreased in premalignant and malignant lesions of the squamous and endocervical mucosa and showed a generally inverse correlation with the expression of p16 at the tissue level. This feedback loop is bypassed via viral E7 interaction and inactivation with pRb, causing p16 to be up regulated which can be detected immunohistochemically. p16 could therefore have a clinical utility as a biomarker because it is a measure of HPV gene expression and activity, rather than solely a detector of viral presence (Stanley 2002). Negri *et al.* in 2003 conducted a study to determine whether immunostaining of p16 is useful in detecting adenocarcinomas of cervix and its precursors in histologic & cytologic routine specimens. A total of 45 patients with glandular lesions including 18 cases of adenocarcinoma in situ (AIS), adenocarcinoma (n=8), endocervical glandular atypia (n=4) and reactive (n=15) lesions were identified. Furthermore, immunocytochemical analysis was performed on 10 Thin Prep Smears with abnormal glandular cells. P16 was detected immunohistochemically on all 26 cases of AIS and adenocarcinoma (100%). Also the immunocytochemical detection on thin prep specimens evidenced a strong expression of p16 in neoplastic endocervical cells. Prior to this study Mc Cluggage *et al.* (2003) investigated the value of p16 immunoreactivity in the distinction between endometrial and endocervical adenocarcinomas. Cases included in this study were endometrial adenocarcinomas of endometrioid type (n=29), and cervical adenocarcinomas of endocervical type (n=23). Twenty-two of 23 endocervical adenocarcinomas showed 100% positive tumor cells. The maximum number of endometrial adenocarcinomas, 9 of 29 showed 21-50% positive tumor cells. They concluded that diffuse strong positivity with p16 suggested an endocervical rather than an endometrial origin of an adenocarcinoma. Endometrial adenocarcinomas are usually positive, but positivity is generally focal and involves less than 50% of cells. Therefore, when there is a morphological doubt then this antibody may be of value as part of a panel for ascertaining the origin of an adenocarcinoma.

Determining the origin of uterine adenocarcinomas can be difficult in biopsy and curettage specimens because the morphologic spectrum of endocervical (ECA) and endometrial adenocarcinomas (EMA) overlap. Ansari Lari *et al.* in 2004 evaluated the utility of immunohistochemistry for P16 in the distinction of ECAs and EMAs. p16 expression was assessed in 24 unequivocal EMA's and 19 unequivocal ECA's and correlated with HPV DNA detected by ISH and PCR. p16 expression was moderate – strong and diffuse in 18 and weak and diffuse in 1 ECA. Fourteen of these were positive for HPV DNA. EMA's displayed weaker staining with patchy distribution and none contained HPV DNA by ISH. Compared with HPV DNA detected by *in situ* hybridization, p16 immunohistochemistry appears to be more sensitive and easier to perform, method for distinguishing ECAs from EMAs. It can be used to assist in the classification of lower uterine segment/endocervical adenocarcinomas of equivocal origin and should be evaluated for its utility in the prospective classification of uterine adenocarcinoma in curettage specimens prior to hysterectomy. Giovanni Negri *et al.* in 2004 evaluated the immunohistochemical expression of p16 as a marker of progression risk in low-grade dysplastic lesions of the cervix uteri. IHC was performed on 32 CIN-I with proven spontaneous regression of lesion in follow up (group A), 31 with progression to CIN-3 (group B) and 33 that were randomly chosen irrespective of the natural history of lesion (Group C). A diffuse staining was detected in 43.8% of CIN-I of group A, 74.2% of group B and 56.3% of group C. Overall 71.4% and 37.8% of p16 negative and diffusely positive CIN-I had regressed, at follow up, where as 26.6% and 62.2% negative and diffusely CIN-I were progressed to CIN-III (p<0.05). Although p16 may be expressed in low grade squamous lesion that undergoes spontaneous regression, in this study CIN-I cases with diffuse p16 staining had a significantly higher tendency to progress to a high grade lesion than p16 negative cases. Therefore, p16 may have the potential to support the interpretation of low grade dysplastic lesions of the cervix uteri. Sahebali *et al.* in 2004 examined the potential of p16 INK4a as a potential biomarker for cervical lesions in a study of liquid based cervical cytology. HPV DNA testing by MY09/MY11 consensus PCR and type specific PCRs and p16INK4a immunocytochemistry on a series of 291 patients selected from routine screening was done. Comparison of the number of p16 immunoreactive cells / 1000 cells exhibited a significantly higher mean count (8.80±1.13) than other cytological groups. The mean count of LSIL (1.09±0.18) was significantly higher than other negative groups. Atypical squamous cells, cannot exclude high grade squamous intraepithelial lesion (ASC-H) and HSIL combined showed a significantly higher mean count (6.46±1.17) than negative

ASC, ASCUS and LSIL. Thus p16 immunocytochemistry can be used as an adjunct to LBC in cervical screening, because it has a good diagnostic accuracy to discriminate HSIL and ASC-H (atypical squamous cells – cannot exclude HSIL) from other lesions. It could be used as a surrogate marker of high risk infections.

Kalof *et al.* in 2005 studied the correlation between p16 immunoexpression, grades of CIN and HPV type in 44 cervical biopsies classified as CIN-I and CIN-II/III. In 22 of 25 CIN-I lesions, p16 immuno expression was confined to lower half of the epithelium with sporadic to focal staining in 11 of 25 cases. In CIN-II/III, 15 of 17 showed diffuse 2/3 to full thickness staining of the epithelial. hr HPV were found in 20 CIN-I lesions and 17 CIN-II/III lesions. Punctate signals were detected in 3 of hr HPV positive CIN-I lesions and 17 of 17 CIN-II/III lesions. They found that p16 immunoexpression and the presence of punctate signal on HPV in situ hybridization correlated with degree of cervical neoplasia ($p<0.001$). Thus both increased p16 immunoexpression and punctate signal correlates with CIN-II/III grade, supporting the use of either, or both tests to confirm CIN-II/III.

P16 can be used as a diagnostic marker along with other well known markers implicated in cervical neoplasia. N Murphy *et al.* in 2005 analysed and compared expression patterns of three potential biomarkers p16, CDC6 and MCM5 and evaluated their use as predictive biomarkers in squamous and glandular pre invasive neoplasia. 20 normal cervical biopsies, in addition to 38 CIN-I, 33 CIN-II, 46 CIN-III, 10 SCC, 19 CGIN and 10 adenocarcinoma were included in the study.. In all normal cases cervical epithelia were not stained. Dysplastic epithelial cells showed p16 staining in 100% of CIN-I, CIN-II, CIN-III, SCC and, adenocarcinoma. Simple linear regression analysis revealed a highly significant linear relation between p16 and increasing grades of squamous dysplasia. Among 3 markers p16 was the most reliable marker of cervical dysplasia. It marked all grades of squamous and glandular lesions of the cervix, and its expression was closely associated with high risk HPV infection. However, the failure of p16 to mark an isolated CIN-III case and staining of glandular mimics as tubo endometrioid metaplasia, may limit its use as a standalone test of cervical dysplasia. Thus a combination of dysplastic markers is suggested in difficult cases.

4.2 MIB-1 as a proliferation marker in cervical neoplasia

MIB-1 (Molecular Immunology Borstel) is an important diagnostic marker for CIN. Gerdes *et al.* in 1990 demonstrated that MIB-1 antibody detects Ki-67 antigen (in paraffin embedded biopsies) in G_1, S, G2 and M phase but is absent in G_0 phase. Baak et al formulated a "Stratification Index" (SI, which indicates, how high Ki-67 positive nuclei are located in the epithelium; the higher the SI, the higher the CIN grade) and the number of Ki-67 nuclei per 100 µm basal membrane (the more Ki-67 nuclei, the higher the grade) to distinguish the three CIN grades at the same time.

Ki-67 is an antigen expressed in proliferating cells (Brown, 2002) that can be detected in formalin fixed tissues using the MIB-1 antibody (Cattoretti et al., 1992). MIB-1 is an important immunocytochemical marker to assess the proliferation and has been suggested as a sensitive biological indicator of progression in CIN lesions by Van Hoven et al., 1997. We performed MIB-1 Immunohistochemistry along with p16 in 63 biopsies of cervical neoplasia and their corresponding Pap smears. MIB-I immunohistochemistry revealed that there was a significant over expression of MIB-1 in different groups and as we move from normal cervical epithelia to varying severity of CINs to carcinoma, the MIB-1 positivity increased. This was found to be statistically significant finding on making a comparison between control versus different groups ($p<0.05$). However, on making an intergroup comparison this was found to be statistically insignificant ($p>0.05$). MIB-1 antibody detects Ki-67 Antigen in G1, S, G2 and M phase but is absent in G0 phase. Therefore, this antibody may be a useful marker of proliferative activity of premalignant and malignant lesions of cervix. In our study we found that as we move from normal to carcinoma group via the varying degrees of CIN, labeling index of positively stained nuclei increased with the severity of CIN to carcinoma group. Review of published literature showed that Goel *et al.* (2005) have also observed similar results. Proliferative index was significantly increased in the carcinoma group in comparison with dysplasia. They showed the following trend for both MIB-1 and PCNA.

Normal < LSIL < HSIL < Carcinoma

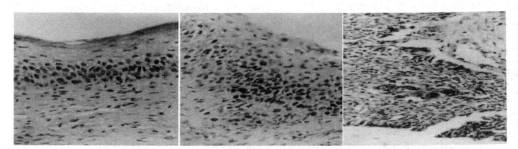

Fig. 2. Images of MIB-1 immunostaining in Cervix tissue biopsz in CINI, CINII, and Carcinoma cervix (40X)

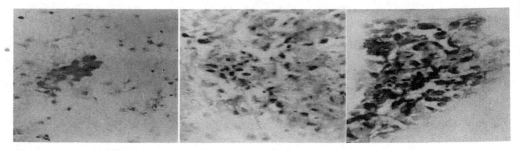

Fig. 3. Images of MIB-1 staining in Pap smears (40X)

In a study by Garzetti *et al.* (1996), MIB-1 immunostaining as an index of cellular proliferations in CIN and micro invasive carcinoma was analysed, with the aim to identify a relationship with the degree of dysplastic lesion and the risk of neoplastic progression. 41 cases of CIN, 23 cases of cervical condyloma, 22 of squamous metaplasia and 10 with micro invasive carcinoma were selected. It was shown that a) positive MIB-1 immunostaining increased progressively from squamous metaplasia to CIN and micro invasive carcinoma, (p<0.001) suggesting that neoplastic proliferation is associated with dysfunctional proliferation of cervical epithelium. b) Considering only CINs the MIB 1 index showed a significant increase with respect to CIN degrees, (p<0.0001). c) That there is a significant correlation between the MIB-1 index and CIN degree but not with respect to HPV DNA presence and d) that MIB-1 immunostaining might be useful for a clinical evaluation of mild and moderate dysplastic lesions. Gorstein et al in 2000 found that in cervical intraepithelial lesions associated with infection by HPV types 16 and 18, the expression of Ki 67 is greater than in lesions unrelated to viral presence.

Prior studies have suggested that Ki-67 (MIB-1) and p16 expression may be preferentially expressed in cervical neoplasia. However, a study conducted by Keating *et al.* in 2001, examined and compared the distribution of staining of these antigens in normal and reactive epithelial changes, diagnostically challenging cases (atypical metaplasia and atrophy) SIL, and high and low risk HPV, type specific SIL. Overall, a histologic diagnosis of SIL correlated strongly with these biomarkers used. Positive scores for Ki-67 and p16 were seen in 68.4% and 100% of LSILS and 94.7% and 100% of HSILs respectively.

P16 INK4a and Ki-67 biomarkers have been evaluated in conventional histopathological sections and more recently on Pap smears. However Akpolat *et al.* in 2004 evaluated the utility of P16 INK4a and Ki-67 staining on cell blocks prepared from residual thin layer cervicovaginal material. Results of cytological based thin prep Pap test were SCC (n=3), HSIL (n=27), LSIL (n=20), ASCUS (n=11), negative for malignancy (n=24). Results of cell blocks preparation were, SCC (n=2), HSIL (n=20), LSIL (n=30), negative for malignancy (n=32). In 62 cases (73%) the diagnosis made using cell blocks were in agreement with thin pap smears. The results indicate that cell blocks represent an additional reliable diagnostic tool in the evaluation of cervical samples[52]. Chisa Aoyama *et al.* in 2005 conducted a study to determine that histologic and immunohistochemical characteristics are useful for distinguishing neoplastic and non-neoplastic lesions. They classified atypical squamous lesion (ASL - a histologic diagnosis of unclear significance in the uterine cervix) (n=37) into neoplastic (n=19) and non-neoplastic (n=18) groups. They chose 7 histologic and IHC indicators to classify ASL. Mitosis, vertical nuclear growth pattern, no perinuclear halo, indistinct cytoplasmic border, primitive cells in the upper third of the squamous layer, p16+ cells in the upper 2/3 of squamous layer and Ki67 positive cells in upper 2/3 of squamous layer were significant indicators for neoplastic ASLs (5 or more of these 7 indicators). Out of 19 ASL, 16 had 5 or more of these indicators. Majority of non-neoplastic ASLs, 16/18 had 2 or fewer indicators.

In a study done by Goel *et al* in 2005, 49 adequate pap smears were stained for MIB-1 and PCNA. Out of 49 cases, 40 cases showed positive immunostaining with MIB-1 and PCNA. Proliferative labelling index of MIB-1 increased with ascending grades of CIN lesions to carcinoma. The highest proliferative index for MIB-1 was observed for the carcinoma group (PCNA-LI 39.200±1.865; MIB-1 LI 35.300±1.888). A significant positive correlation between ascending grades of SIL and LI of markers (r=0.87 for MIB-1 and r=0.88 for PCNA) was seen. This suggests that MIB-1 can be used as an adjunct to cytomorphological interpretation of conventional cervical Pap smear.

	Authors	Year	Number of cases	Results
1	Valasoulis et al	2011	95/LSIL smears	SS=41%;SP=86% PPV=62%;NPV=72%
2	Mendez et al		67/abnormal cytology	35.8% cases positive by p16 and associated with HPV
3	Samir et al	2011	188/pap smears	P16 correlates with increasing CIN grade
4	Balan et al	2010	20/LSIL, HSIL	P16 positive in 68% LSIL;84% CIN2;100%CIN3
5	Schmidt et al	2011	776/ ASCUS, LSIL P16/Ki-67 dual stain cytology	SS=92.2% ASCUS; 94.2% LSIL SP=80.6% ASCUS;68% LSIL
6	Petry et al	2011	425/pap negative; HPV positive P16/Ki-67 dual stain cytology	25.4% positive; SS=91.9% for CIN2; 96.4% for CIN3 SP=82.1% for CIN2; 76.9%for CIN3
7	Alameda et al	2011	109/ frozen sections of ASCUS	SS=82.3%; SP=100%; NPV=94.5%; PPV=100% for HSIL
8	Passamonti et al	2011	91/ ASCUS;60 LSIL;36 ASCH;59 HSIL	46% ASCUS;53% LSIL
9	Srivastava et al	2010	63 /cervical biopsy and pap smears	P16 positive in increasing grades of CIN
10	Bolanca et al	2010	81/ cervical smears	33.3% of HPV positive cases showed p16 positivity
11	Oberg et al	2010	64/ LBC	86% agreement between ProEx C and p16

	Authors	Year	Number of cases	Results
12	Sung et al	2010	105/ ASC-H and ASC-US	P16 correlated significantly with SIL in ASC-H smears
13	Yu et al	2010	63 /cell blocks	HPV L1 and p16 expression increased with severity of cervical lesions
14.	Adamopoulou et al	2009	62 /abnormal pap smears and biopsies	P53, p16 and Bcl-2; SS=83.3%; SP=65.4%
15	Kurshumliu et al	2009	312/ pap smears	36.2% positive for p16
16	Haidopoulos et al	2009	62/abnormal pap smears	SS=100%; SP=76%; PPV=61%; NPV==100%
18	Dray et al.	2005	18/Biopsies 188/Thin pap smears	p16 +ve in HSIL, LSIL. –ve in inflammatory and reactive changes
20	Pientong et al.	2004	165/ pap smear 165/LBC	p16 +ve in 0/30, 21/40, 19/35, 30/30, 30/30 in normal, ASCUS, LSIL, HSIL, Ca.
21	Zielenskii et al.	2002	142/Biopsies of glandular neoplasia	All ACIS and ADCA were HPV positive therefore hr HPV testing is must in cervical cancer screening programme
22	Agoff et al.	2003	569/Biopsies	p16 and Ki67 correlated with cervical neoplasia and HPV
23	Klaes et al.	2002	194/Cervical biopsies	p16 improves the interobserver agreement in diagnosis of CIN

Table 1. Review of literature

LSIL, low grade squamous intraepithelial lesion; SS, sensitivity; SP, specificity; NPV, negative predictive value; PPV, positive predictive value; HPV, human Papilloma virus; CIN, cervical intraepithelial neoplasia; HSIL, high grade squamous intraepithelial lesion; ASCUS, atypical squamous cell of unknown significance; ASCH, atypical squamous cell cannot exclude high grade squamous intraepithelial neoplasia; LBC, liquid base cytology; ACIS, adenocarcinoma in situ; ADCA, adenocarcinoma

5. Conclusion

In a tropical country like India, any perimenopausal women presenting in gynaecological out patient department with any complaint is subjected to a single Pap smear test. However single Pap test is subject to suboptimal sensitivity limited reproducibility and many a times with high rate of false positive and false negative along with equivocal results. To compensate for the aforementioned deficiencies, a screening programme with repeated testing, and follow up of positive cases is warranted. Moreover, colposcopic performed biopsy is directed in any suspicious appearing acetowhite area. This subjects the patient to unnecessary surgical intervention. Therefore, additional diagnostic and prognostic markers for detection of cervical cancers precursors are required which could save the patients from surgical intervention and high screening cost associated with repeated testing.

Also, biomarkers that can help in screening, detection, diagnosis of the disease as well as predict the prognosis can aid the clinicians in correct management of the patients. P16 and MIB-1 are two such candidate markers that fit well in the above mentioned criteria. Through our study we have thus concluded that for LSIL, (because the sensitivity of the p16 marker is 80%), the marker should be evaluated together with MIB-1 or HPV test. For HSIL, the sensitivity and specificity of the p16 marker is 100% and thus it can be used as a stand-alone test. We also recommend that with a careful interpretation of immunostaining with morphological characteristic in the conventional Pap smears, the immunostaining with p16 and MIB-1 markers may be a diagnostic adjunct, reducing the need of tissue biopsy. This is simple, reliable and easily applicable in routine cytosmears. Having said this, there is still need of validation of these markers in a larger cohort and targeted population.

Acknowledgement: The author would like to acknowledge the help of the technical staff of the Department of Pathology, KGMU, Lucknow, India.

6. References

Adamopoulou M, Kalkani E, Charvalos E, Avgoustidis D, Haidopoulos D, Yapijakis C. Comparison of cytology, Colposcopy, HPV typing and biomarker analysis in cervical neoplasia. Anticancer Res. 2009 Aug; 29(8):3401-9.

Agoff et al. p16 expression correlates with degree of cervical neoplasia: A comparison of Ki67 expression and detection of high risk HPV types. Mod. Path. 2003; 16(7): 665-673.

Alameda F, Pijuan L, Lloveras B, Bellosillo B, Larrazabal F, Mancebo G, Muñoz R, Carreras R, Serrano S. The value of p16 in ASCUS cases: a retrospective study using frozen cytologic material. Diagn Cytopathol. 2011 Feb; 39(2):110-4.

AI Nafussi AI, Colquhoun MK. Mild cervical intraepithelial neoplasia (CIN-1) — a histological overdiagnosis. Histopathology 1990; 17:557-61.

Ansari L, Staebler A, Zaino R, Shah K, Ronnett B. Distinction of Endocervical and Endometrial Adenocarcinomas: Immunohistochemical p16 expression correlated with human Papilloma virus (HPV) DNA detection. Am J Surg Pathol 2004 Feb; 28(2): 160-167.

Akpolat I, Smith DA, Ramzy I, Chirala M, Mody DR. The utility of p16INK4a and Ki-67 staining on cell blocks prepared from residual thin-layer cervicovaginal material. Cancer.2004; 102(3):142-9.

Arbyn M, Bergeron C, Klinkhamer P, Martin-Hirsch P, Siebers AG, Bulten J. Liquid compared with conventional cervical cytology. A systematic review and meta-analysis. Obstet Gynecol 2008; 111:167–77.

Baak JPA, Kruse AJ, Robboy SJ, Janssen EAM, van Diermen B, Skaland I. Dynamic behavioural interpretation of cervical intraepithelial neoplasia with molecular markers. J Clin Pathol 2006;59: 1017–1028.

Balan R, Giuşcă S, Căruntu ID, Gheorghiţă V, Neacşu D, Amălinei C. Immunochemical assessment of p16 and HPV L1 capsid protein in cervical squamous intraepithelial lesions. Rev Med Chir Soc Med Nat Iasi. 2010 Oct-Dec; 114(4):1118-24.

Bolanca IK, Sentija K, Simon SK, Kukura V, Vranes J. Estimating clinical outcome of HPV induced cervical lesions by combination of capsid protein L1 and p16INK4a protein detection. Coll Antropol. 2010 Mar;34(1):31-6.

Brinton LA, Reeves WC, Brenes MM et al. The male sexual factor in the etiology of cervical cancer among sexually monogamous women. Int J Cancer 1989;44:199.

Brinton LA, Schairer C, Haenszel W et al. Cigarette smoking and invasive cervical cancer. J Am Med Assoc 1986; 225: 3265

Broder S. From the National Institutes of Health. Rapid Communication — The Bethesda System for Reporting Cervical/Vaginal Cytologic Diagnoses — Report of the 1991 Bethesda Workshop. JAMA 1992;267:1892.

Brown DC, Gatter KC. Ki67 protein: the immaculate deception. Histopathology 2002; 40: 2-11.

Buckley CH, Butler EB, Fox H. Cervical Intraepithelial neoplasia. J Clin Pathol 1982; 35: 1-13.

Butterworth CE, Hatch KD, Macaluso M. Folate deficiency and cervical dysplasia. J Am Med Assoc 1992; 267: 528.

Cattoretti G, Becker MH, Key G. Monoclonal antibodies against recombinant parts of the Ki-67 antigen (MIB-1 and MIB-3) detects proliferating cells in microwave processed formalin fixed paraffin sections. J Pathol 1992, 168: 357-363.

Chisa Aoyama *et al.* Histologic and immunohistochemical characteristics of neoplastic and non-neoplastic subgroups of atypical squamous lesions of the uterine cervix. *Am J Clin Pathol.* 2005, 123: 699-706

Creagh T, Bridger JE, Kupek E, et al. Pathologist variation in reporting cervical borderline epithelial abnormalities and cervical intraepithelial neoplasia. J Clin Pathol 1995; 48:59–60.

Crook T, Wrede D, Vousden KH. p53 point mutation in HPV negative human cervical carcinoma cell lines. Oncogene 1991; 6:873.

Crum CP, Nuovo GJ. Genital papillomaviruses and related neoplasms. New York: Raven Press, 1991, pp. 64-83, 167-185.

Dray M, Russell P, Dalrymple C, Wallman N, Angus G, Leong A, Carter J, Cheerala B. p16 INK 4a as a complementary marker of high? grade intraepithelial lesions of the uterine cervix. I: Experience with squamous lesions in 189 consecutive biopsies. Taylor and Francis Issue 2005 April, 37(2): 112-124.

Fenoglio CM, Galloway DA, Crum CP et al. Herpes simplex virus and cervical neoplasia. In: Fenoglio CM, Wolff M, eds. Progress in surgical pathology. New York: Raven Press, 1982, pp. 45-82.

Garzetti GG, Ciavattini A, De Nictolis M, Lucarini G, Goteri G, Romanini C, Biagini G. MIB 1 immunostaining in cervical intraepithelial neoplasia: prognostic significance in mild and moderate lesions. Gynecol Obstet Invest. 1996; 42(4):261-6.

Gerdes J. Ki-67 and other proliferation markers useful for immunohistological diagnostic and prognostic evaluation in human malignancies. In: Osborn M, editor Seminars in Cancer Biology, 1990; Vol. I: 99-106.

Goel MM, Mehrotra A, Singh U, Gupta HP, Misra JC. MIB-1 and PCNA immunostaining as a diagnostic adjunct to cervical pap smear. Diagnostic Cytopathology 2005; Vol. 32(3).

Gorstein F: Precursor lesions of squamous cell carcinoma of the cervix: are there reliable predictors of biologic behaviour? Hum Pathol 2000; 31: 1339-1340.

Haidopoulos D, Partsinevelos GA, Vlachos GD, Rodolakis A, Markaki S, Voulgaris Z, Diakomanolis E, Antsaklis A. p16 INK4A is a strong biomarker for cervical intraepithelial neoplasia and invasive cervical carcinoma: a reappraisal. Reprod Sci. 2009 Jul; 16(7):685-93. Epub 2009 Apr 16.

Ham J, Dostatni N, Ganthier JM et al. The papillomavirus E2 protein: A factor with many talents. Trends Biochem Sci 1991; 16:440

Jemal A, Bray F, Melissa M, Ferlay J, Ward E, Forman D. Global Cancer Statistics; Ca Can J Clin 2011;61:69–90

Kalof AN, Evans MF, Summons-Arnold L, Beatty BG, Cooper K. p16INK4a immunoexpression and HPV in situ hybridization signal patterns: potential markers of high grade cervical intraepithelial neoplasia. Am J Surg Pathol 2005 May; 29(5): 674-679.

Keating JT, Cviko A, Riethdorf S, Riethdorf L, Quade BJ, Sun D, et al. Ki- 67, Cyclin E, and p16 INK4a Arc complimentary surrogate biomarkers for human papilloma virus – related cervical neoplasia. American Journal of Surgical Pathology 2001; 25:884-91

Klaes R, Friedrich T, Spitkovsky D, Ridder R, Rudy W, Petry U, et al. Over expression of p16 as a specific marker for dysplastic and neoplastic epithelial cells of the cervix uteri. Int J Cancer 2001; 92:276-84.

Klaes, Axel, Friedrich, Tibor, Ridder et al. p16 INK4A Immunohistochemistry improves interobserver agreement in the diagnosis of Cervical Intraepithelial neoplasia. Am. Jour. Surg. Pathol. 2002; 26(11).

Kurshumliu F, Thorns C, Gashi-Luci L. p16INK4A in routine practice as a marker of cervical epithelial neoplasia. Gynecol Oncol. 2009 Oct;115(1):127-31. Epub 2009 Jul 12.

Li J, Li FP, Blot WJ et al. Correlation between cancers of the uterine cervix and penis in China. J Natl Cancer Inst 1982;69: 1063.

Lorincz AT, Reid R, Jenson AB, Greenberg MD, Lancaster W, Kurman RJ et al. Human Papillomavirus infection of the cervix:relative risk associations of 15 common anogenital types. Obstet Gynecol 1992; 79:328-37.

McCluggage WG, Jenkins D. p16 immuno reactivity may assist in the distinction between endometrial and endocervical adenocarcinoma. Int J Gynaecol Pathol 2003 July; 22(3): 231-235.

Meisels A, Morin C. Human Papillomavirus and cancer of the uterine cervix. Gynecol Oncol 1981; 12: 111

Méndez M, Ferrández Izquierdo A. Detection of human papilloma virus (HPV) in liquid-based cervical samples. Correlation with protein p16INK4a expression]. Invest Clin. 2011 Mar; 52(1):3-14.

Munoz N, Bosch FX. Epidemiology of cervical cancer. In: Munoz N, Bosch FX, Jensen OM, eds. Human Papillomavirus and Cervical cancer. Lyon: IARC, 1989; p9

Murphy N, Ring M, Heffron CCBB, King B, Killalea AG, Hughes C, Martin CM, McGuinness E, Sheils O, O'Leary JJ. p16 INK4a, CDC6 and MCM5 : predictive biomarkers in cervical preinvasive neoplasia and cervical cancer. Jour Clin Pathol 2005, 58: 525-534.

Murphy N, Ring M, Killalea AG, Uhlmann V, Donovan MO, Mulcahy F, Turner M, McGuinness E, Griffin M, Martin C, Sheils O, Leary JJ. p16 INK4a as a marker for cervical dyskaryosis : CIN and cGIN in cervical biopsies and thin prep smears. J Clin Pathol 2003; 56: 56-63.

Negri G, Egarter-Vigl E, Kasal A, Romano F, Haitel A, Moan C. p16 INK4a is a useful marker for the diagnosis of Adenocarcinoma of the cervix uteri and its precursors. An Immunohistochemical study with immunocytochemical correlations. Am J Surg Pathol 2003; 27(2): 187-193.

Negri G, Vitadells F, Romano F et al. P16INK 4a expression and progression risk of low grade interepithelial nepolasia of the cervix uteri. Virchows Arch 2004; 445(6), 616-20.

Oberg TN, Kipp BR, Vrana JA, Bartholet MK, Fales CJ, Garcia R, McDonald AN, Rosas BL, Henry MR, Clayton AC. Comparison of p16INK4a and ProEx C immunostaining on cervical ThinPrep cytology and biopsy specimens. Diagn Cytopathol. 2010 Aug; 38(8):564-72.

Passamonti B, Gustinucci D, Recchia P, Bulletti S, Carlani A, Cesarini E, D'Amico MR, D'Angelo V, Di Dato E, Martinelli N, Malaspina M, Spita N. Expression of p16 in abnormal pap-tests as an indicator of CIN2+ lesions: a possible role in the low grade ASC/US and L/SIL (Ig) cytologic lesions for screening prevention of uterine cervical tumours. Pathologica. 2010 Feb; 102(1):6-11.

Petry KU, Schmidt D, Scherbring S, Luyten A, Reinecke-Lüthge A, Bergeron C, Kommoss F, Löning T, Ordi J, Regauer S, Ridder R. Triaging Pap cytology negative, HPV positive cervical cancer screening results with p16/Ki-67 Dual-stained cytology. Gynecol Oncol. 2011 Jun 1; 121(3):505-9.

Pientong et al. Immunocytochemical staining of p16 protein conventional pap test and its association with human papiloma virus infection. Wiley Inter Science. 2004.

Reagen JW, Seidemann IL, Saracusa Y. Cellular morphology of carcinoma in situ and dysplasia or atypical hyperplasia of uterine cervix cancer. Cancer 1953; 6: 224-235

Sahebali J, Depuyde CF, Segars K, Moeneclacy L, Vercecken A, Marck E, Bogers E. p16 as an adjunct Marker in liquid-based cervical cytology. Br J Cancer 2004 108: 871-876.

Samir R, Asplund A, Tot T, Pekar G, Hellberg D. High-Risk HPV Infection and CIN Grade Correlates to the Expression of c-myc, CD4+, FHIT, E-cadherin, Ki-67, and p16INK4a. J Low Genit Tract Dis. 2011 May 7.

Sano T, Oyama T, Kashiwabara K, Fukuda T, Nakajima T. Expression status of p16 protein is associated with HPV oncogenic potential in cervical and genital lesions. Am J Pathol 1998; 153:1741-8.

Scheffner M, Takahashi T, Huibregtse JM et al. Interaction of the human papillomavirus type 16E6 on coprotein with wild-type and mutant human p53 proteins. J Virol 1992; 66:S100.

Scheffner M, Munger K, Huibregtse JM et al. Targeted degradation of the retinoblstoma protein by human Papillomavirus E7-E6 fusion proteins. EMBO J 1992; 11:2425.

Schmidt D, Bergeron C, Denton KJ, Ridder R; European CINtec Cytology Study Group. p16/ki-67 dual-stain cytology in the triage of ASCUS and LSIL papanicolaou cytology: res ults from the European equivocal or mildly abnormal Papanicolaou cytology study. Cancer Cytopathol. 2011 Jun 25;119(3):158-66.

Sherman ME, Schiffman MH, Lorinez AT, Manos MM, Scott DR, Kuman RJ, et al. Toward objective quality assurance in cervical cytopathology: Correlation of cytopathologic diagnoses with detection of high-risk human papillomavirus types. Am J Clin Pathol 1994; 102:182-7.

Solomon D, Davey D, Kurman R, Moriarty A, O'Connor D, Prey M, et al. The 2001 Bethesda System: terminology for reporting results of cervical cytology. JAMA 2002; 287:2114-9.

Spriggs AI, Butler EB, Evans DMD et al. Problems of cell nomenclature in cervical cytology smears. J Clin Pathol 1978; 31: 1226-1227.

Stoler MH, Schiffman M. Atypical Squamous Cells of Undetermined Significance-Low-grade Squamous Intraepithelial Lesion Triage Study (ALTS) Group. Interobserver reproducibility of cervical cytologic and histologic interpretations: realistic estimates from the ASCUS-LSIL Triage Study. JAMA 2001;285:1500-5.

Srivastava S. P16INK4A and MIB-1: an immunohistochemical expression in preneoplasia and neoplasia of the cervix. Indian J Pathol Microbiol. 2010 Jul-Sep;53(3):518-24

Stanley MA. Prognostic factors and new therapeutic approaches to cervical cancer. Virus Res 2002; 89: 241-248.

Sung CO, Kim SR, Oh YL, Song SY. The use of p16(INK4A) immunocytochemistry in "Atypical squamous cells which cannot exclude HSIL" compared with "Atypical squamous cells of undetermined significance" in liquid-based cervical smears. Diagn Cytopathol. 2010 Mar;38(3):168-71.

The 1988 Bethesda System for reporting cervical/ vaginal cytological diagnoses. National Cancer Institute Workshop. JAMA 1989; 262:931-4.

Tomita Y, Shirasawa H, Sekine H, Simizu B. Expression of the human papilloma virus type 6bL2 open reading frame in Escherichia coli : L2 β-galactosidase fusion proteins and their antigenic properties. Virology 1987; 158: 8-14.

Toro de Méndez M, Ferrández Izquierdo A. Detection of human papilloma virus (HPV) in liquid-based cervical samples. Correlation with protein p16INK4a expression].

Tringler B, Gup CJ, Singh M, Groshong S, Shroyer AL, Heinz DE, Shroyer KR. Evaluation of p16 and pRb expression in cervical squamous and glandular neoplasia. Hum Pathol 2004 Jun; 35(6): 689-96

Valasoulis G, Tsoumpou I, Founta C, Kyrgiou M, Dalkalitsis N, Nasioutziki M, Kassanos D, Paraskevaidis E, Karakitsos P. The role of p16(INK4a) immunostaining in the risk assessment of women with LSIL cytology: a prospective pragmatic study. Eur J Gynaecol Oncol. 2011;32(2):150-2.

Van Hoven KH, Rauondetta L, Kovatich AJ, Bibbo L. Quantitative image analysis of MIB I reactivity in inflammatory, hyperplastic and neoplastic endocervical lesions. Int J Gynecol Pathol. 1997; 16: 15-21.

Ward P, Coleman DV, Malcolm ADB. Regulatory Mechanisms of the papillomaviruses. Trends Genet 1989; 5:97.

WHO/ICO Information Centre on HPV and Cervical Cancer. Available from: http://www.who.int/hpv centre. [last cited on 2009 May 5]

World Health Organization. International Classification of Tumors No. 8. In: Ritton G, Christopherson WM Eds. Cytology of the female genital tract. Geneva: World Health Organization, 1973.

Wright TC Jr, Massad LS, Dunton CJ, Spitzer M, Wilkinson EJ, Solomon D, for the 2006 American Society for Colposcopy and Cervical Pathologysponsored consensus conference. 2006 Consensus guidelines for the management of women with abnormal cervical cancer screening tests. Am J Obstet Gynecol. 2007;197(4):346-355.

Yu L, Wang L, Zhong J, Chen S. Diagnostic value of p16INK4A, Ki-67, and human papillomavirus L1 capsid protein immunochemical staining on cell blocks from residual liquid-based gynecologic cytology specimens. Cancer Cytopathol. 2010 Feb 25;118(1):47-55.

Zielinski G, Snijders P, Rozendaal L, Daalmeijer N, Risse E, Voorhorst, Medijiwa N, Linden H, Schipper FA, Runsink AP, Meijer JLM. The presence of high risk-HPV combined with specific p53 and p16 INK4a expression patterns points to high-risk HPV as the main causative agent for adenocarcinoma in situ and adenocarcinoma of the cervix. Journal of Pathology 2003; 201: 535-543.

Zur-Hausen H. Human Papillomaviruses in the pathogenesis of anogenital cancer. Virology 1991;184:9.

Zur- Hausen H. Sexually transmitted diseases and oncogenesis. Abstracts of the Tenth International Meeting of the Society for STD research, Helsinki, Finland,1993, p1.

5

AKNA as Genetic Risk Factor for Cervical Intraepithelial Neoplasia and Cervical Cancer

Kirvis Torres-Poveda, Ana I. Burguete-García, Margarita Bahena-Román, Alfredo Lagunas-Martínez and Vicente Madrid-Marina
Division of Chronic Infection Diseases and Cancer, Centro de Investigación Sobre Enfermedades Infecciosas, Instituto Nacional de Salud Pública, Cuernavaca, Morelos, Mexico

1. Introduction

Cervical cancer (CC) is the second most common cancer among women worldwide. The highest incidence rates of CC are reported in Central and South America, East Africa, South and Southeast Asia and Melanesia. In 2005 there were, according to WHO projections, more than 500, 000 new cases and 274, 000 deaths from CC (WHO, 2007). This is due to the fact that the majority of women in the world do not have access to cervical screening, which can prevent up to 75% of CC cases (Ferlay et al., 2001). Predictions based on the passive growth of the population and the increase in life expectancy say that the expected number of CC cases in 2020 will increase by 40% worldwide, corresponding to 56% in developing countries and 11% in the developed parts of the world (Ferlay et al., 2001).

The development of CC is preceded by a series of cellular abnormalities characterized by cytological and histological changes in cytoplasm maturation and nuclear irregularities. The disease starts as an atypical proliferation of epithelial cells that invade epithelial thickness and degenerate into more serious injuries that invade the stroma. The development of precancerous lesions of the cervix involves several events: exposure to a high-risk Human Papilloma Virus (HR-HPV) causes an initial infection of the squamous epithelium in the transformation zone, followed by morphological and biological alterations of the HPV infected cells (Walboomers et al., 1999).

As the understanding of the natural history of the disease has improved, the classification of these lesions has received different names: PAP I to V, moderate dysplasia, severe carcinoma in situ, cervical intraepithelial neoplasia (CIN) I, II, III, low grade squamous intraepithelial lesions (LSIL) and high grade squamous intraepithelial lesions (HSIL). Microscopically, the evolution of the lesion is characterized by the differentiation of epithelial cells that proliferate and invade the epithelial tissue. Progression is described in terms of increase in the degree of dysplasia (mild, moderate, severe) and carcinoma in situ (El-Ghobashy et al., 2005). Early lesions are now considered manifestations of HPV infection, which are characterized by the presence of nuclear changes and cell proliferation of the epithelium. These cellular abnormalities tend to regress spontaneously, but some of these lesions, particularly those caused by highly oncogenic HPV (16, 18, 31, 33, 35, 45, 56, 58) can modify the thickness of the epithelium and the disease (Nobbenhuis et al., 2001). LSIL and CIN II, have a diploid or polyploid DNA content, which correlates with their tendency to reverse. In contrast, CIN III is often aneuploid, has a greater degree of cellular atypia and is more likely to persist, or progress (Grubisi¢ et al., 2009).

The causal role of HPV in CIN and CC has been firmly established biologically and epidemiologically (Walboomers et al., 1999). Current meta-analysis of the literature shows that the most common HPV types worldwide, in descending order of frequency, are HPV 16, 18, 45, 31, 33, 52, 58 and 35. These are responsible for about 90% of all CC worldwide. The distribution is very similar to that of pre-invasive HSIL (Muñoz et al., 2004; Smith et al., 2007). Since the main route of transmission of genital HPV is sexual, certain patterns of sexual behavior (age of onset of sexual activity, number of sexual partners and sexual behavior of the couple) are associated with an increased risk of infection with genital HPV. But persistent infection with HR-HPV is necessary for carcinogenesis, and cofactors such as multiparity, prolonged use of oral contraceptives, smoking and co-infection with HIV, enhance the progression of infection to cancer (Almonte et al., 2008).

HPVs are species-specific and induce epithelial or fibroepithelial proliferations of benign skin and mucosa in humans and in several animal species. These viruses have a specific tropism for squamous epithelial cells and their production cycle only happens in these cells. HPV infection begins in the basal cells, which are mitotically active. After infection, the virus can lie dormant, replicate and produce infectious particles or become integrated into the cellular genome. Productive infection is divided into several stages depending on the state of differentiation of the epithelial cells. The full cycle that includes viral DNA synthesis, production of viral capsid proteins and assembly of virions, occurs selectively in terminally differentiated keratinocytes (Doorbar, 2006). The basal layer cells proliferate; despite containing HPV DNA, they appear to be very active in the expression of some viral proteins. Apparently, there are cellular factors that negatively regulate viral transcription in these cells.

This regulation is released when the infected cells migrate upward from the epithelium in the granular layer, where they undergo differentiation until they can no longer divide. First, transcription of early and late viral genes is activated in these cells, viral proteins are then synthesized and viral particles are assembled in some superficial cells (Longworth et al., 2004). Although most HPV infections are transient and subclinical, progression is strongly associated with HPV persistence. This process often leads to viral breakthrough in the E1/E2 regions and integration of viral DNA into the cell. The rupture releases the viral E6/E7 promoters and increases the expression of these transformer genes (Moody et al., 2010; Ghittoni et al., 2010). Infection with a HR-HPV acts as a trigger for the cascade of events in which the mechanisms of repair or correction of cell replication, mediated by p53 and retinoblastoma protein (Rb) are altered. Thus, the cell cycle is controlled by the virus, which triggers cellular changes that culminate in the transformation and immortalization of epithelial cells, thus establishing conditions for the start of cancer (Pim & Banks, 2010).

Most HPV infections are transient and intermittent. Epidemiological studies have shown that HPV clearance in healthy, immune competent individuals, takes 8 to 12 months (Woodman et al., 2007). Cohort studies have shown that the continued presence of HR-HPV is necessary for the development, maintenance and progression of HSIL (Ho et al., 1998; Bory et al., 2002; Muñoz et al., 2003; Dalstein et al., 2003; Cuschieri et al., 2005). To establish a persistent infection, HPVs gain access to mitotically active basal-layer keratinocytes, where low-copy replication begins. The viral DNA persists as a nuclear episome in infected cells. In the non-productive stage of infection, HPVs replicate at low copy number in mitotically active basal layer cells within stratified epithelia (Howley, 1996). HPV genomes can persist in proliferating keratinocytes for years or decades. This persistent phase of the viral lifecycle is characterized by detectable levels of viral genome, but the absence of virus production (Stubenrauch & Laimins, 1999), which is most likely a strategy to evade immune surveillance. So, there are two well-defined phenomena in the development of neoplasia - first, there is an alteration in the cellular immune response that allows the persistence of the virus for many years (Walboomers et al., 1999; Nobbenhuis et al., 2001) and second, the phenomena of transformation in epithelial cells produced by oncogenic virus proteins (E6 and E7) through degradation of p53 and pRb (Scheffner et al., 1990; Boyer et al., 1996) and alteration in the expression of proto-oncogenes such as c-Myc and Ras (Marangoz et al., 1999).

Genetic factors intrinsic to the host immune system play a role in susceptibility and/or resistance to the development of CC. Genetic factors associated with CC are polymorphisms in genes coding for tumor necrosis factor alpha (TNF-α) (Wilson et al., 1997; Kirkpatrick et al., 2004), matrix metallopeptidase 1 (Lai et al., 2005), p53 (Dokianakis et al., 2000; Makni et al., 2000), Human Leukocyte Antigen-HLA (Apple et al., 1994; Hildesheim & Wand, 2002) and Fas (Lai et al., 2003). In a complete genome scan in families with more than one child with CC, it was found that in the long arm of chromosome 9 there is a susceptibility locus for the development of cervical carcinoma. In this locus, candidate genes could be potentially involved in genetic predisposition to this disease (Engelmark et al., 2006). Genes such as IKBKAP, PTPN3, TSCOT, AMBP, *akna*, TNFSF15, TNFSF8 and DEC1, have important roles in the development of the immune response.

Among these is the *akna* gene, which is located in band 32 (9q32) on chromosome 9 (HGNC: 24108) and consists of 24 exons (http://www.ncbi.nlm.nih.gov/gene/80709) (Figure 1A). Up to 9 different *akna* transcripts have been identified, resulting from alternative promoter usage, splicing and poly-adenylation. Of the 9 reported transcripts, the 3.1 kb F1 transcript has been functionally tested by our research group (Sims et al., 2005; Perales et al., 2010).

The AKNA protein encoded by this gene is a nuclear protein consisting of 633 amino acids with an approximate weight of 63 kDa. AKNA contains an AT-hook motif of nine amino acids, RTRGRPADS (arginine, threonine, arginine, glycine, arginine, proline, alanine, aspartic acid, serine), which satisfies the consensus requirements of an AT hook DNA-binding motif. The AT-hook is a small motif with a typical sequence signature, in which the tripeptide GRP (Gly[58], Arg[59] and Pro[60]) is the centre of the DNA-binding domain (Siddiqa et al; 2011). Besides, AKNA contains multiple PEST protein-cleavage motifs, which have been shown to target proteins of rapid turnover (Chevallier, 1993; Siddiqa et al., 2001) and has three domains located in the PEST regions: Leu10-Thr43, H149-Ser171, and P616-T63. These regions are sites of protein degradation so AKNA is considered to have a short half-life (Siddiqa et al., 2001) (Figure 1B).

It has been demonstrated that AKNA expresses at least nine transcripts, some of which are expressed in a tissue-specific manner, reflecting its functional diversity. Besides, several of these transcripts are predicted to encode proteins of 155, 137, 100 and 70 kDa. However, only two isoforms (70 and 100 kDa) are expressed in B- and T- cell lines, and both bind to the promoters of the costimulatory molecules of the immune response - CD40 and CD40L (CD40 ligand) (Sims et al; 2005). It has been shown that PEST-dependent cleavage of AKNA is key to its DNA binding function, because in the absence of the PEST-dependent cleavage, the expression of AKNA alone is not sufficient to induce CD40 expression (Sims et al., 2005; Ma et al., 2011).

Fig. 1A. Akna gene localized in chromosome 9Q32

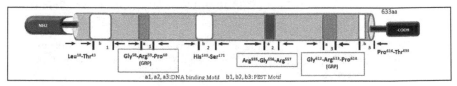

Fig. 1B. Protein at-hook-containning transcription factor

AT-hook proteins have been mainly recognized within the high mobility group A (HMGA) family, which includes chromatin remodeling proteins that coordinate transcriptional complexes to regulate gene expression (Cui & Leng, 2007). However, non-HMG AT-hook proteins have been identified and characterized. Among these, proteins with AT-hook-like motifs (ALMs) have been found to be capable of binding A/T-rich gene targets and regulating their transcription (Senthilkumar & Mishra, 2009; Gordon et al., 2010). In keeping with this notion, the human AKNA gene is a non-HMG transcription factor, which contains N- and C- terminus AT-hook motifs (Siddiqa et al., 2001; Sims et al., 2005). In addition, the AT-hook motif is considered to be the key to upregulation of expression of immune system co-activator molecules, since it has been shown to be directly involved in inducing the expression of molecules belonging to the TNFR family members. AKNA is mainly expressed in secondary lymphoid organs; it is expressed by germinal centre B lymphocytes during B-lymphocyte differentiation and by natural killer (NK) cells and dendritic cells (Siddiqa et al., 2001). AKNA binds the A/T-rich regulatory elements of the CD40 and CD40L promoters (a key receptor-ligand pair that is critical for antigen-dependent B cell development) and induces its expression (Siddiqa et al., 2001).

Thus, in this chapter we will discuss the findings of our research group with respect to the immune response against HPV, CIN and CC, and associated genetic factors of susceptibility to the disease that we found in population studies, particularly the *akna* gene.

2. Immune response against HPV & CIN

In the HPV cycle in keratinocytes, mature virions are shed from the epithelial surface of infected keratinocytes. HPV infection is poorly immunogenic because it is productive (produces no characteristic local inflammation) and during infection there is little presentation of viral antigens to the immune system by professional antigen presenting cells, both locally and systemically (Feller et al., 2010). There is no significant evidence in the literature of inflammation being a risk factor for lesion progression, as observed in several tumor models. On the contrary, the inflammatory infiltrate in patients with established lesions seems to display anti-inflammatory or suppressor characteristics, like the absence of inducible NO synthase (iNOS) expression by macrophages (Mazibrada et al., 2008) and indoleamina2'3'-deoxigenase expression by dendritic cells (Kobayashi et al., 2008).

Interestingly, the number of infiltrating macrophages seems to increase in correlation with lesion grade (Hammes et al., 2007; Kobayashi et al., 2008; Mazibrada et al., 2008). T lymphocytes also infiltrate cervical HPV associated lesions, where the phenotype, abundance and balance between different populations are important in determining the fate of lesions or tumors (Piersma et al., 2004; van der Burg, 2007; Woo et al., 2008). Cellular immune response plays an important role in the control and course of HPV infection; this response varies depending on the degree of injury and the oncogenic potential of HPV. There is evidence that HPV interferes with cell cycle control, secondary to the accumulation of genetic abnormalities; this accounts for viral persistence and the progression of lesions. In many patients secondary factors, such as inadequate immune response, play an important role (Riethmuller & Seilles, 2000).

At the cervical level, after infection of epithelial cells by HPV, a non-specific response is triggered accompanied by chemoattraction of neutrophils, activation of macrophages, activation of NK, natural antibodies and complement system; this is a first defensive barrier of specific immunity. Reticular cells of Langerhans (RCL) and some keratinocytes act as antigen presenting cells (APCs). RCLs are immature dendritic cells of myeloid origin residing in the squamous epithelium, including genital mucosa. RCL recognize the viral particles, capture antigens by micropinocytosis or mannose receptors, process captured proteins and transform them into immunogenic peptides, start the activation process (which includes a surface antigen polypeptide chain with HLA class II, CD40 and B7), migrate to local lymph nodes and present viral peptides to T cells in the context of the major histocompatibility complex (MHC) and costimulatory molecules (CD80, CD86 and CD40). Thus, native lymphocytes are activated and direct their differentiation into effector cells, initiating the antigen-specific immune response (Stern, 2008).

During a primary immune response, depending on the microenvironment and the signals received from certain cytokines, naive CD4+ T cells can differentiate into three to four major subsets of Th cells, with distinct patterns of cytokine secretion that drive different types of immune responses (Seder et al., 2003; Weaver et al., 2006). Briefly, IFN-γ and Interleukin (IL)-12 induce differentiation of Th1 cells to produce more IFN-γ, which enhances the clearance of viruses and the proliferation of specific CD8 cytotoxic T lymphocytes (CTL); on the contrary, if the local context does not express IL-12, it promotes the Th2 pathway which induces activation and expansion of B cells; these evolve, differentiating into plasma cells producing antibodies to viral proteins and inducing the expression of interleukins type IL-4, IL-5, IL-6 and IL-10 (Stern, 2008). The mediators of cell-mediated immunity against HPV infected cells and some tumors are CTL, which eliminate virally infected cells by means of antigen-specific, cell-mediated citotoxicity. Additionally, CD4+ T helper cells participate in the control of these processes. Under certain conditions, the Tumor Growth Factor Beta (TGF-β1) or IL-10 alone drives regulatory CD4+ T cell (Treg) differentiation which regulates immune responses by several distinct mechanisms (Damoiseaux, 2006; Robertson & Hasenkrug, 2006). Thus, together with CTL, the only Th subset that is desirable in pre-malignant CC lesions is that of Th1 cells, as IFN-γ favors immune responses against viral infections.

The natural history of these tumors is long and includes the following steps: infection, persistent infection, viral genome integration into the host cell genome, genomic alterations, immortalization and transformation of epithelial cells. During all these steps, evasion of the immune system is an obligatory feature. The immune evasion mechanisms displayed by the infected cells include absence of cell death (Tindle, 2002), blocking of type I Interferon signaling and reduction of antigen presentation by MHC-I (Ashrafi et al., 2005; Stern, 2008).

2.1 Cellular immune response

Local cellular immune response detected in SIL is characterized by a moderate infiltrate and a decreased inverted Th/Tc (CD4/CD8) ratio, with decreased proliferative capacity (Santin et al., 2001). An imbalance in the pattern of cytokines, as by an increase in type II interleukin (IL-4, IL-10, suppressing the cellular immune response) and a concomitant reduction in interleukin type I [(IL-2, gamma interferon (IFN-γ)], has been reported in women with SIL and CC (Giannini et al., 1998; Clerici et al., 1998; Bor-Ching et al., 2001). Anti-tumor immunity in CC is activated by Th1 cytokines and inhibited by Th2 cytokines. Cytokines such as IL-4, IL-12, IL-10 and/or TGF-β1, produced by various cell types (macrophages, dendritic cells and keratinocytes) have been involved in the suppression of cellular immune response (Giannini et al., 1998).

The pattern of cytokines and their expression in women with CC biopsies has been analyzed. 80% of the tumors express low levels of mRNA of CD4 T lymphocytes and high levels of CD8 CTL. Most tumors express the mRNA of IL-4 and IL-10 and 100% of them express the mRNA of TGF-β1 and IFN-γ. None of the tumors express the IL-12 mRNA, IL-6 mRNA or TNF-α (Alcocer et al., 2006). Immunohistochemical analysis of tumors has demonstrated the presence of IL-10 in tumor cells and cell koilocytes, but not in infiltrating lymphocytes, suggesting that the cells producing IL-10 can be transformed by HPV. On the other hand, a correlation has been found between the immunostaining of IL-10 protein, the level of mRNA IL-10 expression and the supernatants of HPV transformed cell lines expressing IL-10 and TGF-β1 (Alcocer et al., 2006). These findings show a predominant expression of immunosuppressive cytokines, which help to downregulate tumor specific immune response in the tumor microenvironment (Alcocer et al., 2006).

In a recent study, the functionality of peripheral blood T lymphocytes (PBL) and tumor-infiltrating lymphocytes (TIL) of women with SIL and CC was assessed, including proliferation, mRNA expression of IL-2, IFN-γ, IL-4, IL-10 and TGF-β1, as well as the expression of CD3ζ. Using immunohistochemistry, we have seen that TIL is distributed in the stroma more than in epithelium in advanced stages of the disease where CD8 CTL prevail. PBL stimulated with phytohemagglutinin in patients with CC, proliferate less than in SIL patients and healthy subjects. Also, a significant difference is observed in the PBL stimulated with anti-CD3, between patients with CC and healthy subjects. This shows that women with CC have a poor proliferative PBL response and a lack of response to TIL (Díaz et al., 2009).

Antigen recognition of cells with cytotoxic capacity is mediated by receptors. In T cells, these are called T-cell receptors (TCR), which consist of a heterodimer of $\alpha\beta/\gamma\delta$ chains and four ε, δ, γ, ε chains, which together are recognized as TcR-CD3. These receptors are in turn associated with two "zeta" strings which are responsible for the transmission of activation signals within the cell. Zeta chains are also present in the receptor for NK cell antigens. Molecular studies have shown a decreased expression of the dimer formed by the zeta chain, in both T cells and NK cells, contributing to the inefficiency of the effector mechanisms of lymphocytic infiltrate present in the lesions. It has been demonstrated that there is a decreased expression of CD3-zeta chain in patients with CC and CIN (Kono et al., 1996; Shondel et al., 2007), and that suppression *in vivo* can be the result of a circulating factor (Shondel et al., 2007). We found that IL-10 and TGF-ß1 produced by HPV-transformed cells are responsible for CD3-zeta suppression in CC patients (Diaz et al., 2009). These processes are regulated by local factors derived from tumor cells.

As there is an imbalance of cytokines in the microenvironment of these lesions, this affects the transcriptional level. Lymphocytic infiltrate present in the cervical lesions reflects an ineffective immune response (De-Gruijl et al., 1999). A significant correlation between low lymphocyte proliferation and decreased mRNA expression of CD3ζ has been reported in T lymphocytes stimulated with anti-CD3, indicating that T cell function decreases with the progression of CC (Díaz et al., 2009). These changes result in a loss of control of certain HPV 16-18 genes and deregulation in the mechanisms of antigen presentation. Thus, the expression of HLA antigens is reduced or absent (Wang et al., 2002) and there is partial or total absence of Langerhans cells, considered to be they key antigen-presenting immune response against the tumor (Hachisuga et al., 2001).

2.1.1 Role of AKNA in immune response

Proteins containing AT hooks bind A/T-rich DNA through a nine amino-acid motif and are thought to co-regulate transcription by modifying the architecture of DNA, thereby enhancing the accessibility of promoters to transcription factors (Reeves & Nissen, 1990; Friedmann et al., 1993). AKNA is a human AT-hook protein that directly binds the A/T-rich regulatory elements of CD40 and CD40L promoters and coordinately regulates their expression. Consistent with its function, AKNA is a nuclear protein that contains multiple PEST protein-cleavage motifs, which are common in regulatory proteins with high turnover rates (Chevaillier, 1993).

AKNA is mainly expressed by B and T lymphocytes, NK cells and dendritic cells. During B-lymphocyte differentiation, AKNA is mainly expressed by germinal centre B lymphocytes, a stage in which receptor and ligand interactions are crucial for B-lymphocyte maturation (Berek et al., 1991; Liu et al., 1991; Jacob & Kelsoe, 1992; McHeyzer-Williams et al., 1993; Clark & Ledbetter, 1994; MacLennan, 1994; Arpin et al., 1995; Liu et al., 1996). These findings show that an AT-hook molecule can coordinately regulate the expression of a key receptor and its ligand, and point towards a molecular mechanism that explains homotypic cell interactions (Siddiga et al., 2001).

Human AKNA is a transcription factor with N- and C- terminus AT-hook motifs (Siddiga et al., 2001; Sims et al., 2005). Thus, it is possible that AKNA expression plays a role in mechanisms that, if altered, could result in systemic and potentially fatal disorders. Human AKNA is encoded by a single gene located within the FRA9E region of chromosome 9q32 (Sims et al., 2005), a common fragile site (CFS) linked to loss-of-function mutations that often lead to inflammatory and neoplastic diseases (Thye et al., 2003; Landvik et al., 2009; Savas et al., 2009). Based on this reasoning, two independent gene-targeting mouse models were engineered to assess *in vivo* the physiological significance of *akna* gene expression. It was found that the phenotypes resulting from the deletion of the putative C-terminus ALM sequence (AKNA KO) or disruption of AKNA's exon 3 (AKNA KO2) were by and large similar: a) mice died prematurely at neonatal age; b) probable causes of sudden death included acute inflammatory reactions and alveolar destruction; c) triggering of the observed inflammation appeared to be pathogen-induced; d) systemic neutrophil mobilization and alveolar infiltration were routinely observed and; e) concerted activation of neutrophil-specific chemokine, cytokine and proteolytic enzyme expression seemed to be the norm. The central goal of the AKNA function was provided and supports the hypothesis that AKNA expression plays an important role in the mechanisms that regulate the magnitude of inflammatory responses to pathogens (Ma et al., 2011).

It was reported that deletion of murine *akna* gene results in small, frail mice that die suddenly at 10 days of life. Besides, AKNA KO mice present systemic inflammation, predominantly in the lungs, that is accompanied by enhanced leukocyte infiltration (mainly neutrophils) and alveolar destruction. Because AKNA functions as an AT-hook transcription factor, Ma and colleagues investigated expression of genes related to neutrophil function and found a significant enrichment in genes encoding inflammatory cytokines (IL-1β, IFN-γ), inflammatory proteins [neutrophilic granule protein (NGP), cathelin-related antimicrobial peptide (CRAMP), S100A8/9] and proteases (matrix metalloprotease-9; MMP-9), which are implicated in alveolar damage. Given that AKNA deficiency results in an increase of MMP-9, IL-1β, IFN-γ, NGP, and CRAMP S100A9 gene expression, it is possible that AKNA functions as a multi-faceted transcriptional repressor that can coordinately temper pathway-specific gene transcription (Ma et al., 2011; Moliterno & Resar, 2011).

It has been suggested that AKNA´s function is necessary to regulate the magnitude of pathogen-elicited neutrophil activation, proliferation and tissue infiltration by coordinately restricting autocrine/paracrine cytokine and chemokine. This implies that when AKNA is productively expressed, neutrophil reactions are increased to neutralize and destroy pathogens. However, loss of AKNA expression could exacerbate neutrophil activation and cause irreparable tissue damage. This hypothesis is in tune with the enhanced MMP-9, IL-1β, IFN-γ, and NGP expression and the resulting lethal syndrome associated with AKNA deficiency, in which transcriptional repression is seemingly lost (Ma et al., 2011). It is interesting to note that a significantly decreased expression of AKNA in CD4+ T cells has been reported in active patients with Vogt-Koyanagi-Harada Syndrome (VKH, a systemic autoimmune disease). It is unknown whether a decreased AKNA could play a role in VKH syndrome via downregulation of CD40 and CD40L (Mao et al., 2011). In conclusion, it will be interesting to determine if loss-of-function mutations, polymorphisms or epigenetic alterations in *akna* contribute to myeloid function, inflammation and neoplastic transformation.

2.1.2 Evasion of immune surveillance

There is evidence of CIN regression to normal epithelium and it has been suggested that cellular immune response is responsible for HPV clearance (Woo et al., 2010). The cellular immune response mechanisms against HPV are similar to other responses used against viral infection. However, sometimes these mechanisms fail; allow HPV persistence and tumor development. There are several mechanisms for immune response evasion; however, three operate in HPV-infected women: 1) the expression of immunesuppressor cytokines such as, IL-10 and TGF-β1; 2) Fas ligand expression in HPV-transformed cells, and 3) the presence of Treg cells (de Gruijl et al., 1999; Bor-Ching et al., 2001; Alcocer et al., 2006) which are associated with HPV persistence (Molling et al., 2007) and CIN development (Scott et al., 2009). The presence of immune suppressor cytokines in HPV-CIN patient sera has been demonstrated (Clerici et al., 1997).

It is well known that IL-10 and TGF-β1 expression are initiated at the onset of HPV infection and increase as the disease progresses. This allows us to postulate IL-10 as an HPV escape mechanism of the cellular immune response (Bermúdez-Morales et al., 2008). IL-10 and TGF-β1 are expressed in HPV-transformed cells and are induced by HPV E2, E6 and E7 proteins (Bor-Ching et al., 2001; Peralta et al., 2006; Bermúdez et al., 2011).

A susceptibility factor for the development of CC is an alteration of the immune response in patients. This immunosuppression produces a decrease of CTL and NK activation, cells that play an important role in tumor cell elimination. Consequently, HPV-transformed cells do not express antigen presenting molecules, HLA class I and II, an effect mediated by IL-10 and TGF-β1 and by the presence of Treg (Keating et al., 1995; Ploegh, 1998; Nakamura et al., 2007). IL-10 and TGF-β1 also decrease CD3ζ chain expression of the T-cell antigen receptor (Díaz et al., 2009; Patel & Chiplunkar, 2009), which is responsible for antigen recognition. Additionally, it is well known that there are infiltrating T-lymphocytes in CIN and CC, predominantly CD8 CTL; however, these lymphocytes have a low proliferation rate and the absence of costimulatory molecules of cellular immune response. Thus, there is no activation of CTL and no elimination of neoplastic cells (Alcocer et al., 2006; Díaz et al., 2009).

2.2 Humoral immune response

Regarding the humoral response, variability has been reported in antibody titers, regardless of the high levels of circulating immune complexes, especially in those patients with tumors in advanced stages. In early lesions, high levels of IgG antibodies against HPV oncogenic proteins E6/E7 and E4 have been observed, as a result of greater antigenic stimulation, with a decreased IgG1/IgG2 ratio; this is a reflection of the Th1/Th2 imbalance (Matsumoto et al., 1999; Pedroza-Saavedra et al., 2000). In advanced tumors, higher titers of IgA and IgM decrease proportionally as the disease progresses, perhaps due to an impaired immune system (Baay et al., 1997).

3. Genetic susceptibility and AKNA polymorphisms

3.1 Cancer genetic susceptibility

Scientific advances have enabled us to predict susceptibility to developing diverse diseases, including breast, ovarian, and other cancers like CC. With this new knowledge we are able to identify, in a specific population, who are at greater risk of developing a disease.

Cancer is a complex disease that should be considered as a genetic disease. In this respect, genetic factors are not considered to be a direct cause of disease but, in combination with environmental factors, have effects on the resistance or susceptibility to diseases as cancer. Furthermore, cancer risk assessment includes the collection and interpretation of multiple factors that contribute to carcinogenesis. These factors include personal and family health history, reproductive history and hormone use, environmental risk factors such as HPV infection and lifestyle habits associated with cancer risk, as well as any genetic/genomic information (Jenkins, 2009) (Figure 2).

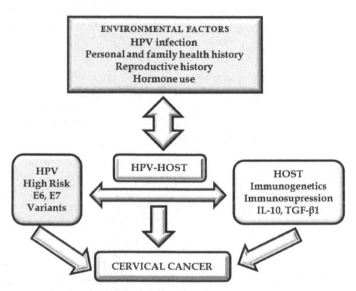

Fig. 2. Factors affecting the persistence of HPV infection and cervical cancer onset

A number of mechanisms leading to cancer have been identified through the discovery of structural alterations of genes called oncogenes and tumour suppressor genes. Somatic and germinal mutations are rare but play a determinant role in the emergence of cancer, while common and frequent variations (polymorphisms) play a role in cancer susceptibility and in the effects of anticancer drugs (efficacy and toxicity)(Robert, 2010; Hildesheim & Wand, 2002).

The susceptibility of a woman to developing CC is largely attributed to the type of HPV infecting the cervix and the persistent HPV infections associated with a high viral load; these are considered to be the major risk factors for persistent cervical lesions (Schlecht et al., 2003; de Araujo Souza & Villa, 2003). Persistence of infections by HR-HPV types is the single greatest risk factor for malignant progression. Although prophylactic vaccines have been developed that target HR-HPV types, there is a continuing need to better understand the virus–host interactions that underlie persistent benign infection and progression to cancer (Bodily & Laimins, 2011). Even though HPV is considered to be a necessary but not sufficient cause for CC, the hereditary component has been reported and several studies indicate that genetic background of the host is important for CC susceptibility and for the carcinogenic process. (de Araujo Souza & Villa, 2003; Hemminki et al., 1999; Magnusson et al., 2000; Wank & Thomssen, 1991).

To study the genetic background and to explore the differences in individual cancer susceptibility, a recent focus of research efforts has been on SNPs (single nucleotide polymorphisms). SNPs are the most common known form of human genetic variation and are defined as stable single-base substitutions with a frequency of greater than 1% in at least one population (Taylor et al., 2001; Meyer et al., 2008). For cancer research, the focus has been on SNPs that alter the function or expression of a gene, to attempt to explain observed associations with a pathogenic mechanism. Indeed, genetic polymorphisms in functionally critical genes such as immune response genes have been suggested as risk factors for the development of a variety of cancers including CC (Taylor et al., 2001; Milam et al., 2007). There are several reports about genes related to immune response (e.g. MHC genes), cytokine genes, genes involved with cancer development (e.g. p53) and CC susceptibility (Glew et al., 1992; Haukin et al., 2002; Sierra-Torres et al., 2003).

Concerning HLA polymorphisms and the risk of CC, discordant results have often been observed among different populations, where the most consistent reports are for the DRB1*13 and DRB1*0603 alleles as a protection factor with OR 0.3-0.4 (Hildesheim & Wand, 2002; Apple et al., 1994; Maciag et al., 2002) and increasing risk for the DQB1*03 and DRB1*1501-DQB1*0602 with OR 2.9 and for the DRB1*0301-DQB1*0201 haplotype, which was associated with a two-fold reduction in risk for transient and persistent HPV infections. DRB1*1102-DQB1*0301 showed a lower risk effect only for persistence. DRB1*1602-DQB1*0502 and DRB1*0807-DQB1*0402 were associated with seven- and three-fold increases in risk for persistence, respectively (Hildesheim & Wand, 2002; Apple et al., 1994; Maciag et al., 2002). More recently, a study of CC cases that were HPV-16 positive and controls, carried out in Mexican population, showed consistent results of association between the HLA-DRB1*15 (OR, 3.9; 95 % CI, 1.6-10.2) and the DRB1*15 DQB1*0602 haplotype (OR, 4.1; 95 % CI, 1.4-12.7) in CC cases, compared with the control group. Also for the HLA class I, the haplotypes HLA-A2-B44-DR4-DQ*0302, HLA-A24-B35-DR16-DQ*0301 and HLA-A2-B40-DR4-DQ*0302 showed a positive association with CC (OR> 1), while HLA-A2-B39-DR4-DQ*0302, HLA-A24-B35-DR4-DQ*0302 and HLA-A68-B40-DR4-DQ*0302 showed a negative association (OR <1) (Hernández et al., 2009).

Also TNF-α has been implicated in direct and indirect control of HPV infection through induction of apoptosis and stimulation of inflammatory responses. Two types of polymorphism have been described in TNF-α gene. One involves several polymorphisms in the TNF-α promoter region, including SNPs at positions -76, -161, -237, -243, -308 (G→A), -375, -568, -572, -575, -857 and -863; the second involves SNPs in DNA microsatellites (Martin et al., 2006). Evidence suggests that the GG genotype of SNP G/T at position -308 in the TNF-α promoter region, has been associated with CC precursor lesions (Kirkpatrick et al., 2004). Other TNF-α SNPs with reported associations with CC include -237, -572, -857 and –863, and haplotype analysis has again strengthened these regions as potential targets for determination of CC susceptibility. The other TNF-α polymorphisms associated with CC are the microsatellite polymorphisms TNF-α2, TNF-α8 and TNF-α11, with significant associations between CIN I and the TNF-α8 allele, CIN III and the TNF-α2 allele in patients who are HPV-18 positive, and TNF-α11 with HPV-16 infection and CIN, in combination with HLA DQB allele (Martin et al., 2006).

Also, p53 polymorphism in codon 72 has been associated with CC. It has been observed that the p53 variants differ biochemically from each other and that the p53-Arg is more susceptible to HPV E6-mediated degradation than the proline variant (Madeleine et al, 2002; Beckman et al., 1994). Furthermore, Storey et al., have shown that individuals who were homozygous for the arginine allele had a seven times higher persistence of HPV associated to CC than heterozygous proline/arginine women (Storey et al., 1998). However, similar analyses performed in other populations did not confirm the association between such polymorphism of the p53 gene and the risk of developing HPV-associated lesions (Minaguchi et al., 1998; Hildesheim et al., 1998). Beside MHC and p53 polymorphism, polymorphisms in cytokine genes can influence immune responses to HPV infection, possibly modifying risks of CC. (Kirkpatrick et al., 2004; Howell & Rose, 2006; Bidwell et al., 1999; Haukin et al., 2002; Stanczuk et al., 2002; Matsumoto et al., 2010).

IL-10 and TGF-β1 polymorphisms have also been reported in diverse populations. G allele of SNP (A/G) at position -1082 in interleukin-10 promoter region, has been associated with high levels of IL-10, which increases with disease severity (p<0.001) (Farzaneh et al., 2006; Singh et al., 2009; Matsumoto et al., 2010). On the other hand, CC patients with -509TT showed marginal low risk for stage I cancer (p = 0.04, OR = 0.95, 95% CI = 0.91-0.99) but -509TT genotype of TGF-β1 was associated with increased risk for stage II cancer (p = 0.07, OR = 3.13, 95% CI = 0.87-11.14) (Singh et al., 2009).

3.2 AKNA polymorphism

A number of genetic susceptibility factors have been proposed, but with the exception of the HLA class II, have not shown consistent results among studies. The first genome-wide linkage scan was performed using 278 affected sib-pairs to identify loci involved in susceptibility to CC. This study found that 9q32 contains the susceptibility locus for CC, and some of these candidate genes are potentially involved in the genetic predisposition to this disease; among these genes is *akna* (Engelmark et al., 2006).

AKNA is a transcriptional factor that is involved in lymphocyte maturation and in the up-regulation of signaling molecules, such as CD40L (Siddiqa et al., 2001; Sims et al., 2005). Even though the precise molecular mechanisms for AKNA function have not been defined, AT-hook transcription factors have emerged as multifaceted regulators that can activate or repress broad A/T-rich gene networks. Thus, alterations of AT-hook genes could affect the transcription of multiple genes causing global cell dysfunction which could mediate DNA bending and chromatin rearrangement (Cairns et al., 1999; Ma et al., 2011).

Although functional data concerning AKNA are scarce, sequences of *akna* genes deposited in the GenBank databases are increasing. SNP analysis in the Genecard site (http://www.genecards.org/), at the 313 SNPs, reported that, for all *akna* genes, only 11 of them are coding non- synonyms. Among all the SNPs reported for *akna*, SNP (rs3748178) involving the transition G/A, at nucleotide 114189600(-) of chromosome 9 (accession no. AK024431) (Ota et al., 2004), appears to be functionally relevant. This mutation produces an R to Q amino acid change at codon 1119 (protein accession NP_110394). Such R/Q mutation occurs at an important AT-hook DNA binding motif within the highly conserved core that has a typical sequence pattern centered around a glycine-arginine-proline (GRP) tripeptide (codons 1118-1120). This short conserved sequence is relevant because it is necessary to bind DNA (Aravind & Landsman, 1998). The importance of producing a GQP core motif, instead of a GRP, is because glutamine lacks the positive charge that arginine has and thus potentially affects its DNA-binding capacity (Reeves & Nissen, 1990; Huth et al., 1997).

The AT-hook motif interacts directly with the minor groove of DNA in AT rich regions. Although some sequence specificity is present in the AT hook itself, and this may affect the main function of this motif (which is to anchor to the proteins in the minor groove of the DNA, near sequences targeted by other regions of the AT hook proteins), it is probably dependent on the spacing between successively interacting AT-hooks and their binding sites may be crucial for conformational changes of the DNA (Huth et al., 1997; Aravind & Landsman, 1998). A SNP at codon 1119 of the *akna* gene, yields a potentially relevant amino acid change (R1119Q) located at the DNA binding AT-hook motif; the AT hook may serve as a contact which affects the specificity and affinity of the DNA binding protein (Figure 3).

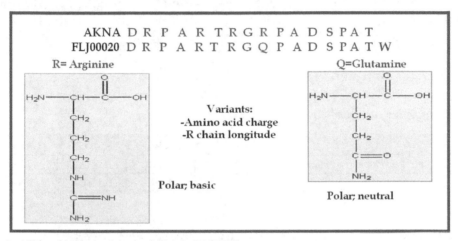

Fig. 3. AT-hook akna polymorphism (rs3748178)

Recently, we examined the frequency of Arginine (R) or Glutamine (Q) 1119 alleles of the *akna* gene in 47 HPV-positive biopsies from Mexican women with cervical lesions, with diagnoses of 21 CIN and 26 CC, as well as in 50 healthy controls without cervical lesions and with HPV-negative status (194 alleles in total). Genomic DNA was amplified by PCR and examined by restriction fragment length polymorphism (RFLP) analysis (Perales et al., 2010). We observed that the frequencies of genotypes in all studied 97 allele pairs were: R/R= 0.597, R/Q= 0.278, Q/Q= 0.123 and the individual frequencies of the R and Q alleles were 0.737 and 0.262, respectively. Q/Q homozygozity was present in 8.33% of healthy controls, 16.67% of CIN and 75% of CC patients. The distribution of the different genotypes in the three study groups showed a statistically significant difference by Fisher's exact test (Perales et al., 2010) (Figure 4).

This study, using a bivariate analysis with a model of multinomial logistic regression, with respective confidence intervals of 95% (IC 95%), showed that these differences were highly significant for the presence of Q/Q in CC (p= 0.01, OR= 3.66, 95%CI: 1.35- 9.94); there was a strong association between the homozygote phenotype Q/Q and the severity of the cervical lesions (Perales et al., 2010). These data support the importance of the genomic region where *akna* is located (Table 1).

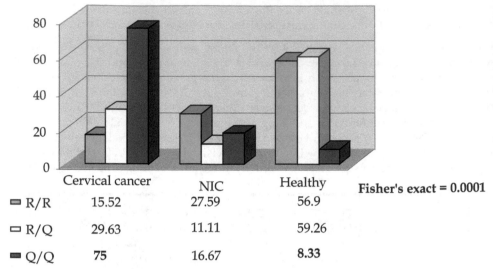

	Cervical cancer	NIC	Healthy	Fisher's exact = 0.0001
R/R	15.52	27.59	56.9	
R/Q	29.63	11.11	59.26	
Q/Q	75	16.67	8.33	

akna genotype distribution, Modified from Perales G, et al,. Biomarkers, 2010

Fig. 4. *akna* genotype distribution

	ORc	CI 95%	P value	ORa	CI95%	P value
CACU	3.66	1.35, 9.94	0.01	4.22	1.3, 13	0.01
NIC	0.6	0.18, 1.93	0.39	0.61	0.18, 2.0	0.42

CI, confidence interval; CIN, cervical intraepithelial neoplasia; Multinomial logistic regression: ORc, unadjusted odds ratio; ORa, adjusted OR by age. Modified from Perales G, et al,. Biomarkers, 2010

Table 1. Risk of Cervical cancer associated with Q/Q *akna* genotype

The AT-hook *akna* motif studied in this work is present in all reported *akna* isoforms except in the C1 and C2 isoforms (Sims et al., 2005). Therefore, the relevance of our observation is valid for the wide range of isoforms potentially expressed by the *akna* gene (Sims et al., 2005). The observed frequencies of *akna* Q/Q genotype in this group of 194 studied chromosomes, and the demonstration of a statistically significant association between a mutation in the *akna* gene (that results in an R-Q amino acid change) and susceptibility to CC, is of high relevance to biological knowledge of the development of CC. Whether this mutation contributes to an alteration of AKNA protein structure and immune function, and to CC development, is under investigation (Perales et al., 2010).

4. Conclusions and perspectives

HPV infection has two different cycles, the productive cycle and the transforming cycle. In the productive cycle of HPV, which occurs in keratinocytes, mature virions are shed from the epithelial surface of infected keratinocytes. HPV infection is poorly immunogenic because it is productive (produces no characteristic local inflammation) and during infection there is little presentation of viral antigens to the immune system by professional antigen-presenting cells, both locally and systemically. In some cases, HPV infection produces a non-inflammatory reaction. There is no adequate cellular immune response, due to excessive secretion of immunosuppressor cytokines, essentially IL-10, which produces a reduction of antigen-presenting and co-stimulatory molecules, and hence no recognition by the surrounding CD8 CTL. The AKNA transcription factor that induces the expression of stimulatory molecules, such as CD40 and CD40L, may play an important role in regulating the immune response against HPV. We found a genetic variant of the *akna* gene that has high frequency in patients with CC. Consequently, we believe that this genetic variant plays an important role, as a risk factor in CC development; the absence of AKNA protein may play an important role in the immune response against HPV. Thus, at the cervical level, after infection of epithelial cells by HPV, a nonspecific response is triggered that is accompanied by chemoattraction of neutrophils. Therefore, this interesting information about *akna*-KO raises important questions and further avenues for biomedical research. For instance, the enrichment of AKNA in wild-type neutrophils, and excess neutrophil counts in the absence of AKNA, suggests that this protein is important in the negative regulation of myeloid differentiation or neutrophil survival. It is of high relevance to determine whether loss-of-function mutations, polymorphisms or epigenetic alterations in *akna* contribute to myeloid diseases, such as myeloproliferative neoplasias, and CC. In conclusion, AKNA appears to be an important genetic factor associated with the progression of CC and genes regulated by this transcription factor could be involved in the resistance to CC progression.

5. References

[1] Alcocer-González, JM., Berumen, J., Tamez-Guerra, R., Bermúdez-Morales, VH., Peralta-Zaragoza, O., Hernández-Pando, R., Moreno, J., Gariglio, P., & Madrid-Marina, (2006). In vivo expression of immunosuppressive cytokines in human papillomavirus-transformed cancer cells. *Viral Immunology*, 19. 3, (July 2006), pp. (481-491), ISSN 1557-8976 (Electronic).

[2] Almonte, M., Albero, G., Molano, M., Carcamo, C., García, PJ., & Pérez G, (2008). Risk factors for human papillomavirus exposure and co-factors for cervical cancer in Latin America and the Caribbean. *Vaccine*, 26. Suppl 11, (August 2008), pp. (L16-36), ISSN 1873-2518 (Electronic).

[3] Apple, RJ., Erlich, HA., Klitz, W., Manos, MM., Becker, TM., & Wheeler, CM, (1994). HLA DR-DQ associations with cervical carcinoma show papilomavirus-type specificity. *Nature Genetics*, 6. 2, (February 1994), pp. (157-162), ISSN 1546-1718 (Electronic).

[4] Aravind, L., & Landsman, D., (1998). AT-hook motifs identified in a wide variety of DNA-binding protein. *Nucleic Acids Research*, 26. 19, (Octubre 1998), pp. (4413-21), ISSN 1362-4962 (Electronic).

[5] Arpin, C., Déchanet, J., Van Kooten, C., Merville, P., Grouard, G., Brière, F., Banchereau, J., Liu, YJ., (1995). Generation of memory B cells and plasma cells in vitro. *Science*, 268. 5211, (May 1995), pp. (720-722), ISSN 1095-9203 (Electronic).

[6] Ashrafi, GH., Haghshenas, MR., Marchetti, B., O'Brien, PM., & Campo, MS., (2005) E5 protein of human papillomavirus type 16 selectively downregulates surface HLA class I. *International Journal of Cancer*, 113. 2, (January 2005), pp. (276-283), ISSN 1097-0215 (Electronic).

[7] Baay, MF., Duk, JM., Groenier, KH., Burger, MP., de Bruijn, HW., Hollema, H., Stolz, E., & Herbrink, P, (1997). Relation between HPV-16 serology and clinico-pathological

data in cervical carcinoma patients: Prognostic value of anti E6 and/or anti E7 antibodies. *Cancer Immunology and Immunotherapy*, 44. 4, (June 1997), pp. (211-213), ISSN 1432-0851 (Electronic).

[8] Beckman, G., Birgander, R., Sjalander, A., Saha, N., Holmberg, PA., Kivela A., & Beckman, L., (1994). Is p53 polymorphism maintained by natural selection? *Human Heredity*, 44. 5, (September 1994), pp. (266–270), ISSN 1423-0062 (Electronic).

[9] Berek, C., Berger, A., & Apel, M., (1991). Maturation of the immune response in germinal centers. *Cell*, 67. 6, (December 1991), pp. (1121-1129), ISSN 1097-4172 (Electronic).

[10] Bermúdez-Morales, VH., Burguete, AI., Gutiérrez, ML., Alcocer-González, JM., & Madrid-Marina, V, (2008). Correlation between IL-10 expression & human papillomavirus infection in cervical cancer. A mechanism for immune response escape. *Cancer Investigation*, 26. 10, (December 2008), pp. (1037–1043), ISSN 1532-4192 (Electronic).

[11] Bermúdez-Morales, VH., Peralta-Zaragoza, O., Moreno, J., Alcocer-González, JM., & Madrid-Marina, V, (2011). IL-10 expression is regulated by HPV E2 protein in cervical cancer cells. *Molecular Medicine Reports*, 4. 2, (Marzo 2011), pp. (369-375), ISSN 1791-3004 (Electronic).

[12] Bidwell, J., Keen, L., Gallagher, G., Kimberly, R., Huizinga, T., McDermott, MF., Oksenberg, J., McNicholl, J., Pociot, F., Hardt, C., & D'Alfonso, S., (1999). Cytokine gene polymorphism in human disease: on line databases. *Genes and Immunity*, 1. 1, (September 1999), pp. (3-19), ISSN 1476-5470 (Electronic).

[13] Bodily, J & Laimins, LA., (2011). Persistence of human papillomavirus infection: keys to malignant progression. Trends in Microbiology, 19. 1, (January 2011), pp. (33-39), ISSN 1878-4380 (Electronic).

[14] Bor-Ching, S., Rong-Hwa, L, Huang-Chung, L., Hong-Nerng, H., Su-Ming, H., & Su-Cheng, H, (2001). Predominant Th2/Tc2 polarity of tumor-infiltrating lymphocytes in human cervical cancer. *Journal of immunology*, 167. 5, (September 2001), pp. (2972-2978), ISSN 1550-6606 (Electronic).

[15] Bory, JP., Cucherousset, J., Lorenzato, M., Gabriel, R., Quereux, C., Birembaut, P., & Clavel, C, (2002). Recurrent human papillomavirus infection detected with the hybrid capture 2 assay selects women with normal cervical smears at risk for developing low grade cervical lesions: a longitudinal study of 3091 women. *International Journal of Cancer*, 102. 5, (December 2002), pp. (519–25), ISSN 1097-0215 (Electronic).

[16] Boyer, SN., Wazer, DE., & Band, V, (1996). E7 protein of human papilloma virus-16 induces degradation of retinoblastoma protein through the ubiquitin–proteasome pathway. *Cancer Research*, 56. 20, (October 1996), pp. (4620–4624), ISSN 1538-7445 (Electronic).

[17] Cairns, B., Schlinchter, A., Erdjument, B., Tempst, P., Kornberg, R., & Winston, F, (1999). Two functionally distinct forms of the RSC nucleosome-remodeling complex, containing essential AT-hook, BAH, and bromodomains. *Molecular Cell*, 4. 5, (November 1999), pp. (715-723), ISSN 1097-4164 (Electronic).

[18] Chevalier, P., (1993). PEST sequences in nuclear proteins. *The International Journal of Biochemistry*, 25. 4, (April 1993), pp. (479-482), ISSN 0020-711X (Print).

[19] Clark, EA & Ledbetter, JA., (1994). How B and T cells talk to each other. *Nature*, 367. 6462, (February 1994), pp. (425-428), ISSN 1476-4687 (Electronic).

[20] Clerici, M., Merola, M., Ferrario, E., Trabattoni, D., Villa, ML., Stefanon, B., Venzon, DJ., Shearer GM., De Palo, G., & Clerici, E, (1997). Cytokine production patterns in cervical intraepithelial neoplasia: association with human papillomavirus infection. *Journal of the National Cancer Institute, 89.* 3, (February 1997), pp. (245–250), ISSN 1460-2105 (Electronic).

[21] Clerici, M., Gene, MS., & Clerici, E, (1998). Cytoquine dysregulation in invasive cervical carcinoma and other human neoplasias: time to consider the Th1/Th2 Paradigm. *Journal of the National Cancer Institute, 90.* 4, (February 1998), pp. (261-263), ISSN 1460-2105 (Electronic).

[22] Cui, T & Leng, F., (2007). Specific recognition of AT-rich DNA sequences by the mammalian high mobility group protein AT-hook 2: a SELEX study. *Biochemistry, 46.* 45, (November 2007), pp. (13059-13066), ISSN 1520-4995 (Electronic).

[23] Cuschieri, KS., Cubie, HA., Whitley, MW., Gilkison, G., Arends, MJ., Graham, C., & McGoogan, E, (2005). Persistent high risk HPV infection associated with development of cervical neoplasia in a prospective population study. *Journal of clinical pathology, 58.* 9, (September 2005), pp. (946–950), ISSN 1472-4146 (Electronic).

[24] Dalstein, V., Riethmuller, D., Prétet, JL., Le, Bail., Carval, K., Sautière, JL., Carbillet, JP., Kantelip, B., Schaal, JP., & Mougin, C., (2003). Persistence and load of high-risk HPV are predictors for development of high-grade cervical lesions: A longitudinal French cohort study. *International Journal of Cancer, 106.* 3, (September 2009), pp. (396-403), ISSN 1097-0215 (Electronic).

[25] Damoiseaux, J., (2006). Regulatory T cells: back to the future. *The Netherlands Journal of Medicine, 64.* 1, (January 2006), pp. (4-9), ISSN 1872-9061 (Electronic).

[26] de Araujo Souza, PS & Villa, LL., (2003). Genetic susceptibility to infection with human papillomavirus and development of cervical cancer in women in Brazil. *Mutation Research, 544.* 2-3, (November 2003), pp. (375–383), ISSN 1873-135X (Electronic).

[27] De-Gruijl, TD., Bontkes, HJ., Peccatori, F., Gallec, MP., Helmerhorst, TJ., Verheijen, RHH., Aarbiou, J., Mulder, WM., Walboomers, JM., Meijer, CJ., van de Vange, N., & Scheper, RJ., (1999). Expression of CD3-zeta on T-cell sin primary cervical carcinoma and in metastasis-positive and negative pelvic lymph nodes. *British Journal of Cancer, 79.* 7-8, (March 1999), pp. (1127-1132), ISSN 1532-1827 (Electronic).

[28] Díaz-Benítez, CE., Navarro-Fuentes, KR., Flores-Sosa, JA., Juárez-Díaz, J., Uribe-Salas, FJ., Román-Basaure, E., González-Mena, LE., Alonso de Ruíz, P., López-Estrada, G., Lagunas-Martínez, A., Bermúdez-Morales, VH., Alcocer-González, JM., Martínez-Barnetche, J., Hernández-Pando, R., Rosenstein, Y., Moreno, J., & Madrid-Marina, V., (2009). CD3ζ expression and T cell proliferation are inhibited by TGF-β1 and IL-10 in cervical cancer patients. *Journal of Clinical Immunology, 29.* 4, (July 2009), pp. (532–544), ISSN 1573-2592 (Electronic).

[29] Dokianakis, D., & Spandidos, D., (2000). p53 Codon 72 polymorphism as a risk factor in the development of HPV-associated cervical cancer. *Molecular Cell Biology Research Communications, 3.* 2, (February 2000), pp. (111–114), ISSN 1522-4732 (Electronic).

[30] Doorbar J., (2006). Molecular biology of human papillomavirus infection and cervical cancer. *Clinical Science, 110.* 5, (May 2006), pp. (525-541), ISSN 0009-9287 (Print).

[31] El-Ghobashy, AA., Shaaban, AM., Herod, J., & Herrington, CS., (2005). The pathology and management of endocervical glandular neoplasia. *International Journal of Gynecological Cancer*, 15. 4, (July 2005), pp. (583-592), ISSN 1525-1438 (Electronic).

[32] Engelmark, MT., Ivansson, EL., Magnusson, JJ., Gustavsson, IM., Beskow, AH., Magnusson, PK., & Gyllensten, UB., (2006). Identification of susceptibility loci for cervical carcinoma by genome scan of affected sib-pairs. *Human Molecular Genetics*, 15. 22, (November 2006), pp. (3351-3360), ISSN 1460-2083 (Electronic).

[33] Farzaneh, F., Roberts, S., Mandal, D., Ollier, B., Winters, U., Kitchener, HC., & Brabin, L., (2006). The IL-10 -1082G polymorphism is associated with clearance of HPV infection. *British journal of obstetrics and gynaecology*, 113. 8, (August 2006), pp. (961-964), ISSN 0306-5456 (Print).

[34] Feller, L., Wood, NH., Khammissa, RA., Chikte, UM., Meyerov, R., & Lemmer, J., (2010). HPV modulation of host immune responses. *South African Dental Association*, 65. 6, (July 2010), pp. (266-268), ISSN 1029-4864 (Print).

[35] Ferlay, J., Bray, F., Pisani, P., & Parkin, DM., (2001). GLOBOCAN 2000: Cancer Incidence, Mortality and Prevalence Worldwide. Versión 1.0. IARC Cancer Base N.o 5. Lyon: IARC Press. Retrieved from http://www-dep.iarc.fr/globocan/globocan.htm.

[36] Friedmann, M., Holth, LT., Zoghbi, HY., & Reeves, R., (1993). Organization, inducible-expression and chromosome localization of the human HMG-I(Y) nonhistone protein gene. *Nucleic Acids Research*, 21. 18, (September 1993), pp. (4259-4267), ISSN 1362-4962 (Electronic).

[37] Giannini, SL., Al-Saleh, W., Piron, H., Jacobs, N., Doyen, J., Boniver, J., & Delvenne, P., (1998). Cytokine expresion in squamous intraepithelial lesions of the uterine cervix : implications for the generation of local inmunosuppression. *Clinical and Experimental Immunology*, 113. 2, (August 1998), pp. (183-189), ISSN 1365-2249 (Electronic).

[38] Ghittoni, R., Accardi, R., Hasan, U., Gheit, T., Sylla, B., & Tommasino, M., (2010). The biological properties of E6 and E7 oncoproteins from human Papillomaviruses. *Virus Genes*, 40. 1, (February 2010), pp. (1-13), ISSN 1572-994X (Electronic).

[39] Glew, SS., Stern, PL., Davidson JA., & Dyer, PA., (1992). HLA antigens and cervical carcinoma. *Nature* 356. 6364, (March 1992), pp. (22), ISSN 1476-4687 (Electronic).

[40] Gordon, BR., Li, Y., Wang, L., Sintsova, A., van Bakel, H., Tian, S., Navarre, WW., Xia, B., & Liu, J., (2010). Lsr2 is a nucleoid-associated protein that targets AT-rich sequences and virulence genes in Mycobacterium tuberculosis. *Proceedings of the National Academy of Sciences of the United States of America*, 107. 11, (March 2010), pp. (5154-5159), ISSN 1091-6490 (Electronic.

[41] Grubisi¢, G., Klari¢, P., Jokanovi¢, L., Soljaci¢-Vranes, H., Grbavac, I., & Bolanca, I., (2009). Diagnostic approach for precancerous and early invasive cancerous lesions of the uterine cervix. *Collegium antropologicum*, 33. 4, (December 2009), pp. (1431-1436), ISSN 0350-6134 (Print).

[42] Hachisuga, T., Fukuda, K., & Kawarabayashi, T., (2001). Local immune response in squamous cell carcinoma of the uterine cervix. *Gynecologic and Obstetric Investigation*, 52. 1, (July 2001), pp. (3-8), ISSN 1423-002X (Electronic).

[43] Hammes, LS., Tekmal, RR., Naud, P., Edelweiss, MI., Kirma, N., Valente, PT., Syrjänen, KJ., & Cunha-Filho, JS., (2007). Macrophages, inflammation and risk of cervical

intraepithelial neoplasia (CIN) progression-clinicopathological correlation. *Gynecologic Oncology*, 105. 1, (April 2007), pp. (157-165), ISSN 1095-6859 (Electronic).

[44] Haukin, N., Bidwell, JL., Smith, AJ., Keen, LJ., Gallagher, G., Kimberly, R., Huizinga, T., McDermott, MF., Oksenberg, J., McNicholl, J., Pociot, F., Hardt, C., & D'Alfonso, S., (2002). Cytokine gene polymorphism in human disease: on line databases. Supplement 2. *Genes and Inmunity*, 3. 6, (September 2002), pp. (313-330), ISSN 1476-5470 (Electronic).

[45] Hemminki, K., Dong, C., & Vaittinen, P., (1999). Familial risks in cervical cancer: is there a hereditary component? *International Journal of Cancer*, 82. 6, (September 1999), pp. (775–781), ISSN), ISSN 1097-0215 (Electronic).

[46] Hernández-Hernández, DM,. Cerda-Flores, RM., Juárez-Cedillo, T., Granados-Arriola, J., Vargas-Alarcón, G., Apresa-García, T., Alvarado-Cabrero, I., García-Carrancá, A., Salcedo-Vargas, M., & Mohar-Betancourt, A., (2009). Human leukocyte antigens I and II haplotypes associated with human papillomavirus 16-positive invasive cervical cancer in Mexican women. *International Journal of Gynecological Cancer*, 19. 6, (August 2009), pp. (1099-10106), ISSN 1525-1438 (Electronic).

[47] Hildesheim, A., Schiffman, M., Brinton, LA., Fraumeni, JF Jr., Herrero, R., Bratti, MC., Schwartz, P., Mortel, R., Barnes, W., Greenberg, M., Mcgowan, L., Scott, DR., Martin, M., Herrera, JE., & Carrington, M., (1998). p53 polymorphism and risk of cervical cancer. *Nature*, 396. 6711, (December 1998), pp. (531–532), ISSN 1476-4687 (Electronic).

[48] Hildesheim, A., & Wand, SS., (2002). Host and viral genetics & risk of cervical cancer: a review. Virus Research, 89. 2, (November 2002), pp. (229-240), ISSN 1872-7492 (Electronic).

[49] Ho, GY., Bierman, R., Beardsley, L., Chang, CJ., & Burk, RD., (1998). Natural history of cervicovaginal papilloma virus infection in young women. *The New England Journal of Medicine*, 338. 7, (February 1998), pp. (423–428), ISSN 1533-4406 (Electronic).

[50] Howley, PM., (1996) Virology. 2nd edition ed. Philadelphia, Pa.: Lippincott-Raven Publishers; Papillomavirinae: the viruses and their replication, pp. (2045-2076).

[51] Howell, WM & Rose-Zerilli, MJ., (2006). Interleukin-10 polymorphisms, cancer susceptibility and prognosis. *Familial Cancer*, 5. 2, (July 2006), pp. (143-149), ISSN 1573-7292 (Electronic).

[52] Huth, J., Bewley, C., Nissen, M., Evans, J., Reeves, R., Gronenborn, A., & Clore, G., (1997). The solution structure of an HMG-I(Y)-DNA complex defines a new architectural minor groove binding motif. *Nature Structural Biology*, 4. 8, (August 1997), pp. (657-665), ISSN 1072-8368 (Print).

[53] Jacob, J & Kelsoe, G., (1992). In situ studies of the primary immune response to (4-hydroxy-3-nitrophenyl)acetyl. II. A common clonal origin for periarteriolar lymphoid sheath-associated foci and germinal centers. , 176. 3, (September 1992), pp. (679-687), ISSN1540-9538 (Electronic).

[54] Jenkins JF. (Ed(s)). *(2009). Consensus Panel on Genetic/Genomic Nursing Competencies, Essentials of genetic and genomic nursing: competencies, curricula guidelines, and outcome indicators*, American Nurses Association, Silver Spring, MD (Ed 2), ISBN-13: 978-1-55810-263-7 ISBN-10: 1-55810: American Nurses Association.

[55] Keating, PJ., Cromme, FV., Duggan-Keen, M., Snijers, PJ., Walboomers, JM., Hunter, RD., Dyer, PA., & Stern, PL., (1995). Frequency of down-regulation of individual

HLA-A and B alleles in cervical carcinomas in relation to TAP-1 expression. *British Journal of Cancer*, 72. 2, (August 1995), pp. (405-411), ISSN 1532-1827 (Electronic).

[56] Kirkpatrick, A., Bidwell, J., van den Brule, A., Meiler, C., Pawad, J., & Glew S., (2004). TNFα polymorphism frequencies in HPV-associated cervical dysplasia. *Gynecologic Oncology*, 92. 2, (February 2004), pp. (675–679), ISSN 1095-6859 (Electronic).

[57] Kobayashi, A., Weinberg, V., Darragh, T., & Smith-McCunne, K., (2008). Evolving immunosuppressive microenvironment during human cervical carcinogenesis. *Mucosal Immunology*, 1. 5, (September 2008), pp. (412-420), ISSN 1935-3456 (Electronic).

[58] Kono, K., Ressing, ME., Brandt, RM., Melief, CJ., Potkul, RK., Andersson, B., Petersson, M., Kast, WM., & Kiessling, R., (1996). Decreased expression of signal-transducing zeta chain in peripheral T cells and natural killer cells in patients with cervical cancer. *Clinical Cancer Research*, 2. 11, (November 1996), pp. (1825-1828), ISSN 1078-0432 (Print).

[59] Landvik, NE., Hart, K., Skaug, V., Stangeland, LB., Haugen, A., & Zienolddiny, S., (2009). A specific interleukin-1B haplotype correlates with high levels of IL1B mRNA in the lung and increased risk of non-small cell lung cancer. *Carcinogenesis*, 30. 7, (July 2009), pp. (1186-1192), ISSN 1460-2180 (Electronic).

[60] Lai, H., Sytwu, H., Sun, C., Yu, C., Liu, H., Chang, C., & Chu, T., (2003). Single Nucleotide Polymorphism at Fas promoter is associates with Cervical Carcinogenesis. *International Journal of Cancer*, 103. 2, (January 2003), pp. (221–225), ISSN 1097-0215 (Electronic).

[61] Lai, HC., Chu, CM., Lin, YW., Chang, CC., Nieh, S., Yu, MH., & Chu, TY., (2005). Matrix metalloproteinase 1 gene polymorphism as a prognostic predictor of invasive cervical cancer. *Gynecologic Oncology*, 96. 2, (February 2005), pp. (314-319), ISSN 1095-6859 (Electronic).

[62] Liu, YJ., Zhang, J., Lane, PJ., Chan, EY., & MacLennan, IC., (1991). Sites of specific B cell activation in primary and secondary responses to T cell-dependent and T cell-independent antigens. *European Journal of Immunology*, 21. 12, (December 1991), pp. (2951-2962), ISSN 1521-4141 (Electronic).

[63] Liu, YJ., Malisan, F., de Bouteiller, O., Guret, C., Lebecque, S., Banchereau, J., Mills, FC., Max, EE., & Martinez-Valdez, H., (1996). Within germinal centers, isotype switching of immunoglobulin genes occurs after the onset of somatic mutation. *Immunity*, 4. 3, (March 1996), pp. (241-250), ISSN1097-4180 (Electronic).

[64] Longworth, MS., & Laimins, LA., (2004). Pathogenesis of human papillomaviruses in differentiating epithelia. *Microbiology and Molecular Biology Reviews*, 68. 2, (June 2004), pp. (362-372), ISSN 1098-5557 (Electronic).

[65] Ma, W., Ortiz-Quintero, B., Rangel, R., McKeller, MR., Herrera-Rodríguez, S., Castillo, EF., Schluns, KS., Hall, M., Zhang, H., Suh, WK., Okada, H., Mak, TW., Zhou, Y., Blackburn, MR., & Martínez-Valdez, H., (2011). Coordinate activation of inflammatory gene networks, alveolar destruction and neonatal death in AKNA deficient mice. *Cell Research*, Epub ahead of print (May 2011), pp. (1-14), ISSN 1748-7838 (Electronic).

[66] Maciag, PC., Schlecht, NF., Souza, PS., Rohan, TE., Franco, EL., & Villa, LL., (2002). Polymorphisms of the human leukocyte antigen *DRB1* and *DQB1* genes and the

natural history of human papillomavirus infection. *Journal of infectious diseases*, 186. 2, (July 2002), pp. (164–172), ISSN 1537-6613 (Electronic).

[67] MacLennan, IC., (1994). Germinal centers. *Annual Review of Immunology*, 12, (July 1994), pp. (117-139), ISSN 1545-3278 (Electronic).

[68] Madeleine, MM., Brumback, B., Cushing-Haugen, KL., Schwartz, SM., Daling, JR., Smith, AG., Nelson, JL., Porter, P., Shera, KA., McDougall, JK., & Galloway, DA., (2002). Human leukocyte antigen class II and cervical cancer risk: a population-based study. *Journal of infectious diseases*, 186. 11, (December 2002), pp. (1565–1574), ISSN 1537-6613 (Electronic).

[69] Makni, H., Franco, E., Kaiano, J., Villa, L., Labrecque, S., Dudley, R., Storey, A., & Matlashewski, G., (2000). p53 polymorphism in codon 72 and risk of human papilloma virus-induced cervical cancer: effect of inter-laboratory variation. *International Journal of Cancer*, 87. 4, (August 2000), pp. (528–533), ISSN ISSN 1097-0215 (Electronic).

[70] Magnusson, PK., Lichtenstein, P., & Gyllensten UB., (2000). Heritability of cervical tumours. *International Journal of Cancer*, 88. 5, (December 2000), pp. (698–701), ISSN ISSN 1097-0215 (Electronic)

[71] Mao, L., Yang, P., Hou, S., Li, F., & Kijlstra, A., (2011). Label-Free proteomics reveals decreased expression of CD18 and AKNA in peripheral CD4+ T Cells from patients with Vogt-Koyanagi-Harada Syndrome. *Plos One*, 6. 1, (January 2011), pp. (e14616), ISSN 1932-6203 (Electronic).

[72] Marangoz, S & Güllü, IH., (1999). Expression of ras, c-myc, and p53 proteins in cervical intraepithelial neoplasia. *Cancer*, 85. 12, (June 1999), pp. (2668-2669), ISSN 1097-0142 (Electronic).

[73] Martin, CM., Astbury, K., & O'Leary, JJ., (2006). Molecular profiling of cervical neoplasia. *Expert Review of Molecular Diagnostics*, 6. 2, (March 2006), pp. (217-229), ISSN 1744-8352 (Electronic)

[74] Matsumoto, K., Yoshikawa, H., Yasugi, T., Nakagawa, S., Kawana, K., Nozawa, S., Hoshiai, H., Shiromizu, K., Kanda, T., & Taketani, Y., (1999). Balance of IgG subclasses toward human papillomatype 16 (HPV 16) L1-capside is a possible predictor for the regression of HPV16-positivecervical intraepithelial neoplasia. *Biochemical and Biophysical Research Communications*, 258. 1, (April 1999), pp. (128-131), ISSN 1090-2104 (Electronic).

[75] Matsumoto, K., Oki, A., Satoh, T., Okada, S., Minaguchi, T., Onuki, M., Ochi, H., Nakao, S., Sakurai, M., Abe, A., Hamada, H., & Yoshikawa, H., (2010). Interleukin-10 -1082 gene polymorphism and susceptibility to cervical cancer among Japanese women. *Japanese Journal of Clinical Oncology*, 40. 11, (November 2010), pp. (1113-1116), ISSN 1465-3621 (Electronic).

[76] Mazibrada, J., Rittà, M., Mondini, M., De Andrea, M., Azzimonti, B., Borgogna, C., Ciotti, M., Orlando, A., Surico, N., Chiusa, L., Landolfo, M., & Gariglio, M., (2008). Interaction between inflammation and angiogenesis during different stages of cervical carcinogenesis. *Gynecologic Oncology*, 108. 1, (January 2008), pp. (112-120), ISSN 1095-6859 (Electronic).

[77] McHeyzer-Williams, MG., McLean, MJ., Lalor, PA., & Nossal, GJ., (1993). Antigen-driven B cell differentiation in vivo. *The Journal of experimental medicine*, 178. 1, (July 1993), pp. (295-307), ISSN 1540-9538 (Electronic).

[78] Meyer, LA., Westin, SN., Lu, KH, & Milam, MR., (2008). Genetic polymorphisms and endometrial cancer risk. *Expert Review of Anticancer Therapy*, 8. 7, (July 2008), pp. (1159-1167), ISSN 1744-8328 (Electronic).

[79] Milam, MR., Gu, J., Yang, H., Celestino, J., Wu, W., Horwitz, IB., Lacour, RA., Westin, SN., Gershenson, DM., Wu, X., & Lu, KH., (2007). STK15 F31I polymorphism is associated with increased uterine cancer risk: a pilot study. *Gynecologic Oncology*, 107. 1, (October 2007), pp. (71-74), ISSN 1095-6859 (Electronic).

[80] Minaguchi, T., Kanamori, Y., Matsushima, M., Yoshikawa, H., Taketani, Y., & Nakamura, Y., (1998). No evidence of correlation between polymorphism at codon 72 of p53 and risk of cervical cancer in Japanese patients with human papillomavirus 16/18 infection. *Cancer Research*, 58. 20, (October 1998), pp. (4585–4586), ISSN 1538-7445 (Electronic).

[81] Moliterno, AR & Resar, LMS., (2011). AKNA: Another AT-hook transcription factor "hooking-up" with inflamation. *Cell Research*, Epub ahead of print (June 2011), pp. (1-3), ISSN 1748-7838 (Electronic).

[82] Molling, JW., de Gruijl, TD., Glim, J., Moreno, M., Rozendaal, L., Meijer, CJ., van den Eertwegh, AJ., Scheper, RJ., von Blomberg, ME., & Bontkes, HJ., (2007). CD4(+)CD25hi regulatory T-cell frequency correlates with persistence of human papillomavirus type 16 and T helper cell responses in patients with cervical intraepithelial neoplasia. *International Journal of Cancer*, 121. 8, (October 2007), pp. (1749-1755), ISSN 1097-0215 (Electronic).

[83] Moody, CA & Laimins, LA., (2010). Human papillomavirus oncoproteins: Pathways to transformation. *Nature reviews Cancer*, 10. 8, (August 2010), pp. (550-560), ISSN 1474-1768 (Electronic)

[84] Muñoz, N., Bosch, FX., de Sanjosé, S., Herrero, R., Castellsagué, X., Shah, KV., Snijders, PJ., Meijer, CJ., & International Agency for Research on Cancer Multicenter Cervical Cancer Study Group., (2003). Epidemiologic classification of human papillomavirus types associated with cervical cancer. *The New England Journal of Medicine*, 348. 6, (February 2003), pp. (518–527), ISSN 1533-4406 (Electronic).

[85] Muñoz, N., Bosch, FX., Castellsagué, X., Díaz, M., de Sanjosé, S., Hammouda, D., Shah, KV., & Meijer, CJ., (2004). Against which human papillomavirus types shall we vaccinate and screen? The international perspective. *International Journal of Cancer*, 111. 2, (August 2004), pp. (278-285), ISSN 1097-0215 (Electronic).

[86] Nakamura, T., Shima, T., Saeki, A., Hidaka, T., Nakashima, A., Takikawa, O., & Saito, S., (2007). Expression of indoleamine 2, 3-dioxygenase and the recruitment of Foxp3-expressing regulatory T cells in the development and progression of uterine cervical cancer. *Cancer Science*, 98. 6, (June 2007), pp. (874-881), ISSN 1349-7006 (Electronic).

[87] Nobbenhuis, ME., Helmerhorst, TJ., van den Brule, AJ., Rozendaal, L., Woorhorst, FJ., Bezemer, PD., Verheijen, RH., & Meijer, CJ., (2001). Cytological regression and clearance of high-risk human papilomavirus in women with an abnormal cervical smear. *Lancet*, 358. 9295, (November 2001), pp. (1782-1783), ISSN 1474-547X (Electronic).

[88] Ota T, Suzuki Y, Nishikawa T, Otsuki T, Sugiyama T, Irie R, Wakamatsu A, Hayashi K, Sato H, Nagai K, & others., (2004). Complete sequencing and characterization of

21,243 full-length human cDNAs. *Nature Genetics*, 36. 1, (January 2004), pp. (40–45), ISSN 1546-1718 (Electronic).

[89] Patel, S & Chiplunkar, S., (2009). Host immune responses to cervical cancer. *Current Opinion in Obstetrics & Gynecology*, 21. 1, (February 2009), pp. (54-59), ISSN 1473-656X (Electronic).

[90] Pedroza-Saavedra, A., Cruz, A., Esquivel, F., De La Torre, F., Berumen, J., Gariglio, P., & Gutiérrez, L., (2000). High prevalence of serum antibodies to Ras and type 16 E4 proteins of human papillomavirus in patients with precancerous lesions of the uterine cervix. *Archives of Virology*, 145. 3, (July 2000), pp. (603-623), ISSN 1432-8798 (Electronic).

[91] Perales, G., Burguete-García, AI., Dimas, J., Bahena-Román, M., Bermúdez-Morales, VH., Moreno, J & Madrid-Marina., (2010). V. A polymorphism in the AT-hook motif of the transcriptional regulador AKNA is a risk factor for cervical cancer. *Biomarkers*, 15. 5, (August 2010), pp. (470-474), ISSN 1366-5804 (Electronic).

[92] Peralta-Zaragoza O, Bermúdez-Morales V, Gutiérrez-Xicotencatl L, Alcocer-González, J., Recillas-Targa, F., & Madrid-Marina, V., (2006). E6 and E7 oncoproteins from human papillomavirus type 16 induce activation of human transforming growth factor beta1 promoter throughout Sp1 recognition sequence. *Viral Immunology*, 19. 3, (July 2006), pp. (468-480), ISSN 1557-8976 (Electronic).

[93] Piersma, SJ., Jordanova, ES., van Poelgeest, MI., Kwappenberg, KM., van der Hulst, JM., Drijfhout, JW., Melief, CJ., Kenter, GG., Fleuren, GJ., Offringa, R., & van der Burg, SH., (2004). High number of intraepithelial CD8+ tumor-infiltrating lymphocytes is associated with the absence of lymph node metastases in patients with large early-stage cervical cancer. *Cancer Research*, 67. 1, (January 2007), pp. (354-361), ISSN 1538-7445 (Electronic).

[94] Pim D & Banks L., (2010). Interaction of viral oncoproteins with cellular target molecules: infection with high-risk vs low-risk human papillomaviruses. *Acta Pathologica, Microbiologica, et Immunologica Scandinavica*, 118, 6-7, (June 2010), pp. (471-493), ISSN 1600-0463 (Electronic).

[95] Ploegh, HL., (1998). Viral strategies of immune evasion. *Science*, 280. 5361, (April 1998), pp. (248-253), ISSN 1095-9203 (Electronic).

[96] Reeves, R & Nissen, MS., (1990). The A.T-DNA-binding domain of mammalian high mobility group I chromosomal proteins. A novel peptide motif for recognizing DNA structure. *The Journal of Biological Chemistry*, 265. 15, (May 1990), pp. (8573-8582), ISSN 1083-351X (Electronic).

[97] Reeves, R., (2010). Nuclear functions of the HMG proteins. BBA *International Journal of Biochemistry and Biophysics*, 1799. 1-2, (January 2010), pp. (3-14), ISSN 0006-3002 (Print).

[98] Riethmuller, D & Seilles, E., (2000). Inmunity of the female genital tract mucosa and mechanisms of papillomavirus evasion. *Journal de Gynécologie, Obstétrique et Biologie de la Reproduction*, 29. 8, (July 2000), pp. (729-740), ISSN 0368-2315 (Print).

[99] Robert, J., (2010). Gene polymorphisms. Bull Cancer, 97. 11, (November 2010), pp. (1253-1264), ISSN 1769-6917 (Electronic).

[100] Robertson, SJ & Hasenkrug, KJ., (2006). The role of virus-induced regulatory T cells in immunopathology. *Springer Seminars in Immunopathology*, 28. 1, (August 2006), pp. (51-62), ISSN 1432-2196 (Electronic).

[101] Santin, AD., Ravaggi, A., Bellone, S., Pecorelli, S., Cannon, M., Parham, GP., & Hermonat PL., (2001). Tumor-infiltring lymphocytes contain higher numbers of type1 cytoquine expressors and DR+ T cells compared with Lymphocytes from tumor drainig lymph nodes and peripheral blood in patients with cancer of the uterine cervix. *Gynecologic Oncology*, 81. 3, (June 2001), pp. (424-432), ISSN 1095-6859 (Electronic).

[102] Savas, S & Liu, G., (2009). Genetic variations as cancer prognostic markers: review and update. *Human Mutations*, 30. 10, (October 2009), pp. (1369-1377), ISSN 1098-1004 (Electronic).

[103] Scheffner, M., Werness, BA., Huibregtse, JM., Levine, AJ., & Howley, PM., (1990). The E6 oncoprotein encoded by human papillomavirus types 16 and 18 promotes the degradation of p53. *Cell*, 63. 6, (December 1990), pp. (1129–1136), ISSN 1097-4172 (Electronic).

[104] Schlecht, NF., Trevisan, A., Duarte-Franco, E., Rohan, TE., Ferenczy, A., Villa, LL., & Franco, EL., (2003). Viral load as a predictor of the risk of cervical intraepithelial neoplasia, *International Journal of Cancer*, 103. 4, (February 2003), pp. (519–524), ISSN 1097-0215 (Electronic).

[105] Scott, ME., Ma, Y., Kuzmich, L., & Moscicki, AB., (2009). Diminished IFN-gamma and IL-10 and elevated Foxp3 mRNA expression in the cervix are associated with CIN 2 or 3. *International Journal of Cancer*, 124. 6, (March 2009), pp. (1379-1383), ISSN 1097-0215 (Electronic).

[106] Seder, RA & Ahmed, R., (2003). Similarities and differences in CD4+ and CD8+ effector and memory T cell generation. *Nature Immunology*, 4. 9, (September 2003), pp. (835-842), ISSN 1529-2916 (Electronic).

[107] Senthilkumar, R & Mishra, RK., (2009). Novel motifs distinguish multiple homologues of Polycomb in vertebrates: expansion and diversification of the epigenetic toolkit. *BMC Genomics*, 10. (November 2009), pp. (549), ISSN 1471-2164 (Electronic).

[108] Shondel, SM., Helm, CW., Gercel-Taylor, C., & Taylor, DD., (2007). Differential expression of T-cell CD3-zeta chains in patients with cervical dysplasia before and after treatment. *International Journal of Gynecological Cancer*, 17. 6, (November 2007), pp. (1278-1282), ISSN 1525-1438 (Electronic).

[109] Siddiqa, A., Sims-Mourtada, JC., Guzmán-Rojas, L, Rangel, R., Guret, C., Madrid-Marina, V., Sun, Y., & Martínez-Valdez, H., (2001). Regulation of CD40 and CD40 ligand by the AT-hook transcription factor AKNA. *Nature*, 410. 6826, (March 2001), pp. (383–387), ISSN 1476-4687 (Electronic).

[110] Sierra-Torres, CH., Au, WW., Arrastia, CD., Cajas-Salazar, N., Robazetti, SC., Payne, DA., & Tyring, SK., (2003). Polymorphisms for chemical metabolizing genes and risk for cervical neoplasia. *Environmental and Molecular Mutagenesis*, 41. 1, (July 2003), pp. (69–76), ISSN 1098-2280 (Electronic).

[111] Singh, H., Jain, M., & Mittal, B., (2009). Role of TGF-beta1 (-509C>T) promoter polymorphism in susceptibility to cervical cancer. *Oncology Research*, 18. 1, (July 2009), pp. (41-45), ISSN 0965-0407 (Print).

[112] Sims-Mourtada, JC., Bruce, S., Mckeller, MR., Rangel, R., Guzmán-Rojas, L., Cain, K., López, C., Zimonjic, DB., Popescu, NC., Gordon, J., Wilkinson, MF., & Martínez-Valdez, H., (2005). The Human AKNA Gene Expresses Multiple Transcripts and Protein Isoforms as a Result of Alternative Promoter Usage, Splicing, and

Polyadenylation. *DNA and Cell Biology*, 24. 5, (May 2005), pp. (325-338), ISSN 1557-7430 (Electronic).

[113] Smith, JS., Lindsay, L., Hoots, B., Keys, J., Franceschi, S., Winer, R., & Clifford, GM., (2007). Human papillomavirus type distribution in invasive cervical cancer and high-grade cervical lesions: a meta-analysis update. *International Journal of Cancer*, 121. 3, (August 2007), pp. (621-632), ISSN 1097-0215 (Electronic).

[114] Stanczuk, GA., Tswana, SA., Bergstrom, S., & Sibanda, EN., (2002). Polymorphism in codons 10 and 25 of the transforming growth factor-beta 1 (TGF-beta1) gene in patients with invasive squamous cell carcinoma of the uterine cervix. *European Journal of Immunogenetics*, 29. 5, (October 2002), pp. (417-421), ISSN 1365-2370 (Electronic).

[115] Stern, PT., (2008). Natural immune control of HPV infection, In: *Vaccines for the prevention of cervical cancer.* Peter L Stern, Henry C Kitchener, pp. (57-65), Oxford University Press, ISBN 978-0-19-954345-8, New York.

[116] Storey, A., Thomas, M., Kalita, A., Harwood, C., Gardiol, D., Mantovani, F., Breuer, J., Leigh, IM., Matlashewski, G., & Banks, L., (1998). Role of a p53 polymorphism in the development of human papillomavirus-associated cancer. *Nature*, 393. 6682, (May 1998), pp. (229–234), ISSN 1476-4687 (Electronic).

[117] Stubenrauch, F & Laimins, LA., (1999). Human papillomavirus life cycle: active and latent phases. *Seminars in Cancer Biology*, 9. 6, (December 1999), pp. (379–386), ISSN 1096-3650 (Electronic).

[118] Taylor, JG., Choi, EH., Foster, CB., & Chanock, SJ., (2001). Using genetic variation to study human disease. *Trends in Molecular Medicine*, 7. 11, (November 2001), pp. (507-512), ISSN 1471-499X (Electronic).

[119] Thye, T., Burchard, GD., Nilius, M., Muller-Myhsok, B., & Horstmann, RD., (2003). Genomewide linkage analysis identifies polymorphism in the human interferon-gamma receptor affecting Helicobacter pylori infection. *American Journal of Human Genetics*, 72. 2, (February 2003), pp. (448-453), 1537-6605 (Electronic).

[120] Tindle, RW., (2002). Immune evasion in human papillomavirus-associated cervical cancer. *Nature Reviews Cancer*, 2. 1, (January 2002), pp. (59-65), ISSN 1474-1768 (Electronic).

[121] van der Burg, SH., Piersma, SJ., de Jong, A., van der Hulst, JM., Kwappenberg, KM., van den Hende, M., Welters, MJ., Van Rood, JJ., Fleuren, GJ., Melief, CJ., Kenter, GG., & Offringa, R., (2007). Association of cervical cancer with the presence of CD4+ regulatory T cells specific for human papillomavirus antigens. *Proceedings of the National Academy of Sciences of the United States of America*, 104. 29, (July 2007), pp. (12087-12092), ISSN 1091-6490 (Electronic).

[122] Walboomers, JM., Jacobs, MV., Manos, MM., Bosch, FX., Kummer, JA., Shah, KV., Snijders, PJ., Peto, J., Meijer, CJ., & Muñoz, N., (1999). Human papilomavirus is a necessary cause of invasive cervical cancer worldwide. *The Journal of Pathology*, 189. 1, (September 1999), pp. (12-19), ISSN 1096-9896 (Electronic).

[123] Wang, SS., Hildesheim, A., Gao, X., Schiffman, M., Herrero, R., Bratti, MC., Sherman, ME., Barnes, WA., Greenberg, MD., McGowan, L., Mortel, R., Schwartz, PE., Zaino, RJ., Glass, AG., Burk, RD., Karacki, P., & Carrington, M., (2002). Human leukocyte antigen class I alleles and cervical neoplasia no heterozygote advantage. *Cancer*

Epidemiology, Biomarkers and Prevention, 11. 4, (April 2002), pp. (419-420), ISSN 1538-7755 (Electronic).

[124] Wank, R & Thomssen, C., (1991). High risk of squamous cell carcinoma of the cervix for women with HLA-DQw3. *Nature*, 352. 6337, (August 1991), pp. (723–725), ISSN 1476-4687 (Electronic).

[125] Weaver, CT., Harrington, LE., Mangan, PR., Gavrieli, M., & Murphy, KM., (2006). Th17: an effector CD4 T cell lineage with regulatory T cell ties. *Immunity*, 24. 6, (June 2006), pp. (677-688), ISSN 1097-4180 (Electronic).

[126] World Health Organization., 2007. *Integración de la atención sanitaria para la salud sexual y reproductiva y las enfermedades crónicas. Control integral del CaCU.* Guía de prácticas esenciales, pp. 281.

[127] Wilson, A., Symons, J., McDowell, T., McDevitt, HO., & Duff, G., (1997). Effects of a polymorphism in the human tumor necrosis factor a promoter on transcriptional activation. *Proceedings of the National Academy of Sciences of the United States of America*, 94. 7, (April 1997), pp. (3195–3199), ISSN 1091-6490 (Electronic).

[128] Woo, YL., Sterling, J., Damay, I., Coleman, N., Crawford, R., van der Burg, SH., & Stanley, M., (2008) Characterising the local immune responses in cervical intraepithelial neoplasia: a crosssectional and longitudinal analysis. *British Journal of Obstetrics and Gynaecology*, 115. 13, (December 2008), pp. (1616-1621), ISSN 1471-0528 (Electronic).

[129] Woo, YL., van den Hende, M., Sterling, JC., Coleman, N., Crawford, RA., Kwappenberg, KM., Stanley, MA., & van der Burg, SH., (2010). A prospective study on the natural course of low-grade squamous intraepithelial lesions and the presence of HPV16 E2-, E6- and E7-specific T-cell responses. *International Journal of Cancer*, 126. 1, (January 2010), pp. (133-141), ISSN 1097-0215 (Electronic).

[130] Woodman, CB., Collins, SI., & Young, LS., (2007). The natural history of cervical HPV infection: unresolved issues. *Nature Reviews Cancer*, 7. 1, (January 2007), pp. (11-22), ISSN 1474-1768 (Electronic).

6

The Role of the Pap Smear Diagnosis: Atypical Glandular Cells (AGC)

Chiung-Ru Lai[1,2], Chih-Yi Hsu[1,2] and Anna Fen-Yau Li [1,2]
[1]Department of Pathology and Laboratory Medicine,
Taipei Veterans General Hospital,
[2]School of Medicine, National Yang-Ming University, Taipei,
Taiwan

1. Introduction

Glandular lesions of the female genital tract have always been a challenge for pathologists. The precise cytological diagnosis of these lesions is difficult because of their inherent complexity, as well as the lack of experience of many cytopathologists in this field.

The term atypical glandular cells of undetermined significance (AGUS) was first introduced at the 1988 Bethesda Conference (National Cancer Institute Workshop, 1989) and defined as morphologic changes in glandular cells beyond those suggestive a benign reactive process, but insufficient for the interpretation of adenocarcinoma. In the 2001 Bethesda System (TBS 2001) (Solomon, 2002), the term has been changed to better reflect current knowledge and understanding of glandular neoplasia. The category has been defined and renamed "atypical glandular cells" (AGC), with the subclassifications "not otherwise specified" (AGC-NOS) and "favor neoplastic" (AGC-FN). The cell type of origin, endocervical or endometrial, should be addressed whenever possible. Adenocarcinoma in situ has been separated as another distinct category of diagnosis.

In 2006, the American Society for Colposcopy and Cervical Pathology (ASCCP) released consensus guidelines for the management guidelines for the management of AGC (Wright, 2007). The guidelines emphasized combined colposcopy and endocervical sampling was recommended for all women across all subcategories of AGC, with the addition of endometrial sampling for women over 35. So, since 2006, more comprehensive evaluations were applied for these women. Recent studies concerning the follow-up outcomes of AGC revealed more patients with precancerous or malignant diseases of different sites ranging from the exo-cervix, endocervix, endometrium, fallopian tube, ovary and even extra-genital organs (Behtash, 2007; Duska, 1998; Jeng, 2003; Koonings, 2001; Lai, 2008; Manetta, 1999; Mood, 2006; Soofer, 2000). Since the introduction of Pap smear screening, the incidence of cervical squamous cell carcinoma has been dramatically declined but the relative incidence of glandular cancer has been increased. However, the sensitivity of detecting cervical glandular precancerous or cancer lesions is much less than that of the squamous lesions making cervical glandular cancer prevention remains a challenge and problem to be solved

(Koss, 1989; Wingo, 2003). So, our ability to recognize and diagnose AGC-NOS or AGC-FN is very important. After correct triage of patients with AGC Pap smears, early treatment of these lesions may be achieved. The protective effects of cytologic screening for glandular lesions can be then improved.

In our previous study (Lai, 2008), it supported the view that a diagnosis of AGC is clinically significant by the 2001 Bethesda System, especially the AGC-FN category. The subclassification of AGC is important and demanded in the diagnosis of Pap smears. Addressing the cell origin of endometrium, although being found no statistically significant difference, it showed a more common significant pathology outcome. Since then, we still followed the 2001 Bethesda System to subclassify and address the cell origin in AGC Pap smears. As to management protocol, we strongly recommend following the ASCCP consensus guidelines. In the current retrospective study of 9 years experience, histological follow-up results obtained and paired to the corresponding cytology interpretation, and the results further enhanced the importance of the role of the Pap smear diagnosis of AGC in screening and diagnosing the precancerous and cancer lesions.

2. Materials and methods

A retrospective review of the archives of the Department of Pathology, Taipei Veterans General Hospital, from January 2002 to December 2010 identified 234 smears diagnosed as AGC with at least 6 months follow-up. All of the Pap smears since January 2002 were diagnosed and classified according to the 2001 Bethesda System criteria at the time of diagnosis. If cellular findings suggestive of endometrial glandular or stromal cells were noted, the description of "endometrial origin" would be made in the space of "educational notes and suggestion" in the cytologic report. An adequate evaluation for AGC Pap smears suggested by the ASCCP included a colposcopy with or without cervical biopsy, endocervical curettage and an endometrial sampling, especially in those patients in whom endometrial origin was addressed in the Pap test. In addition, those patients who received other diagnostic or treatment procedures such as conization, loop electrosurgical excision procedure (LEEP) or hysterectomy were also included in this study. The most abnormal histology was considered to be the outcome. Patients who failed to receive the management described even with multiple repeated pap smears were excluded in the evaluation.

The clinical information of patient, such as age, menopausal status, hormonal replacement therapy status, tamoxifen use status, and presence of abnormal bleeding were collected from medical record. Pathology findings of endometrial biopsy were categorized as benign, precursors (high grade squamous intra-epithelial lesion, endocervical adenocarcinoma in situ, endometrial atypical complex hyperplasia), and malignant. The precursors and malignant pathology results are defined as abnormal pathology. Based on cyto-histological and available clinical data, we made meticulous description of the cytological findings including atypical glandular cells themselves and the background pattern and statistical analyses on the different subclassifications of AGC by using Chi-square test and multivariate logistic regression. A P value <0.05 was considered to be statistically significant.

3. Results

3.1 Pathology results

From a total of 228,451 cervicovaginal cytologic specimens within a 9-year period from January 2002 to December 2010, a total of 234 (0.1%) AGC Pap smears were identified. The age distribution ranged from 27 to 93 years (median 49). All were conventional Pap smears and primarily carried out for cervical cancer screening. 190 of 234 (81%) cases with adequate histologic evaluation were included in this study. (Table 1)

AGC subtype	Number	Histologic follow-up (%)
AGC-NOS	197	157 (80%)
AGC-FN	37	33 (89%)
Total	234	190 (81%)

AGC: atypical glandular cells
AGC-NOS: atypical glandular cells, not otherwise specified
AGC-FN: atypical glandular cells, favor neoplastic

Table 1. Histologic follow-up rates by AGC subtype

Adequate initial evaluation for AGC Pap smears suggested by the ASCCP included a colposcopy with or without cervical biopsy, endocervical curettage and an endometrial sampling, especially in those patients endometrial origin was addressed in the Pap test. Abnormal histology of precursors and invasive lesions were found in 76 patients (40%) (Table 2) Final pathology results included 37 endometrial adenocarcinomas, 6 endocervical adenocarcinomas, 1 cervical squamous cell carcinoma, 1 endometrial malignant mixed Mullerian tumor (MMMT), 3 ovarian carcinomas, 3 colon-rectal adenocarcinomas, 1 fallopian tube adenocarcinoma, 4 endocervical adenocarcinoma in situ, 2 endocervical glandular dysplasia, 6 high grade squamous intraepithelial lesion (HSIL), and 11 endometrial complex hyperplasia. Invasive diseases, accounting for 28% (53 of 190) were much more common than precursors, 12% (23 of 190). All of the patients with significant pathology received definitive treatment, including complete staging surgery for those harboring invasive neoplastic diseases.

There were 83 smears sub-classified as AGC-NOS; 75 as AGC-NOS, endometrial origin (EM); 21 as AGC-N, endometrial origin (EM); 11 as AGC-N. The subgroup of AGC-N, EM had the highest rate of abnormal pathology, followed by AGC-NOS, EM, AGC-N and AGC-NOS; 18 of 21 (86%), 30 of 75 (40%), 4 of 11 (36%) and 24 of 83 (29%), respectively. The difference was significant. ($P<0.001$) Women with AGC-N were more likely to have significant pathology (22 in 32 (69%)) compared with those with AGC-NOS (54 in 158 (34%)). It was statistically significant. ($P<0.001$) The endometrial origin addressed cases had more abnormal pathology results than those not being addressed, 48 of 96 (50%) v.s. 28 of 94 (30%). ($P=0.004$) (Table 3)

Histologic results	Cases (%)
Benign	114 (60%)
Abnormal	76 (40%)
Invasive lesions	53 (28%)
Cervical cancer	7
Adenocarcinoma	6
Squamous cell carcinoma	1
Endometrial cancer	38
Adenocarcinoma	37
Malignant mixed Müllerian tumor	1
Extra-uterine malignancies	8
Ovary carcinoma	3
Rectal cancer	4
Tubal carcinoma	1
Precursor lesions	23 (12%)
Endocervical glandular dysplasia	2
Endocervical adenocarcinoma in situ	4
High grade squamous intraepithelial lesion	6
Endometrial complex hyperplasia (including atypical)	11

Table 2. Final Histologic results of 190 patients with AGC Pap smears

	N	Benign	Abnormal	P
Diagnostic category				
AGC-NOS, EM	75	45 (60%)	30 (40%)	<0.001[1]
AGC-NOS	83	59 (71%)	24 (29%)	
AGC-FN, EM	21	3 (14%)	18 (86%)	
AGC-FN	11	7 (64%)	4 (36%)	
EM				
Address	96	48 (50%)	48 (50%)	0.004[1]
Not address	94	66 (70%)	28 (30%)	
Favor neoplastic				
Yes	32	10 (31%)	22(69%)	<0.001[1]
No	158	104 (66%)	54 (34%)	
Total cases	190	114 (60%)	76 (40%)	

[1]chi-square test
EM: endometrial origin

Table 3. Abnormal histologic results in different AGC subtypes

3.2 Cytologic findings and differential diagnoses

Degenerative atypical endometrial glandular cells admixed with endometrial debris indicated endometrial origin, The endometrial debris distributed along the smearing direction (Figure 1-3) was characterized by watery diatheses, foamy histiocytes, degenerative necrotic debris and phagocytosis (Figure 4). Some or all of the above findings were observed in 1 fallopian tube adenocarcinoma , 25 endometrial adenocarcinomas and 1 MMMT but none of the cervical lesions and other extra-uterine cancers.

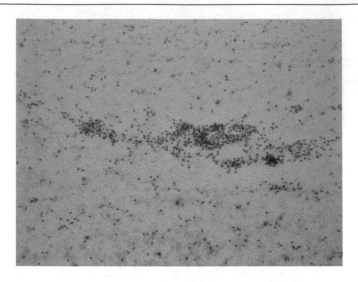

Fig. 1. The endometrial debris distributed along the smearing direction in the background. (Papanicolaou stain, 100x)

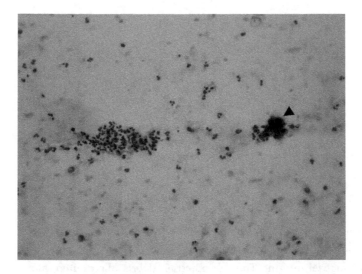

Fig. 2. Degenerative atypical endometrial glandular cells, favor neoplastic (arrow head), admixed with endometrial debris. The final pathology turned out to be an endometrioid adenocarcinoma, grade II. (Papanicolaou stain, 200x)

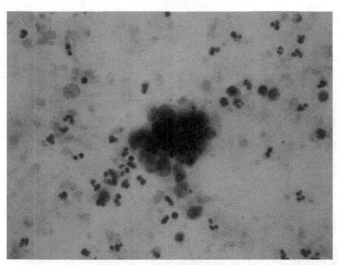

Fig. 3. The higher magnification showed tight cluster of atypical endometrial glandular cells with degeneration, high N/C ratio, three-dimensional structure, and small faint nucleoli. These features fall short of diagnosing an adenocarcinoma directly, either in quantity or quality. (Papanicolaou stain, 400x)

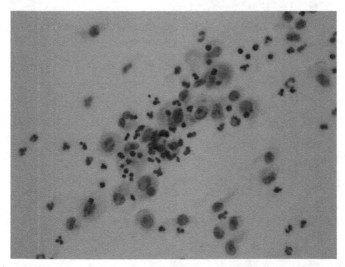

Fig. 4. The endometrial debris was characterized by watery diatheses, foamy histiocytes, degenerative necrotic debris and phagocytosis. It was very specific for endometrial lesions. (Papanicolaou stain, 400x)

The distribution pattern, along the smearing direction, was very characteristic for this kind of debris indicating shedding from endometrium instead of endocervix. However, it would disappear in the fluid-based preparations. In addition, mucin substance was always absent in the endometrial debris. On the contrary, we noticed that a characteristic finding consisting of necrosis and a mucinous background resembling the pattern seen in ileal conduit urine was an indicator suggestive of endocervical adenocarcinoma (Figure 5-8).

This feature was seen in 3 of the 6 endocervical adenocarcinomas but none of the other cancers.

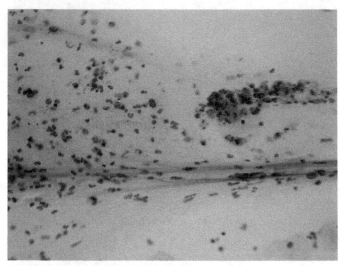

Fig. 5. Background of mucin streaks admixed with necrotic mucous cells resembling those of an ileal conduit urine specimen. (Papanicolaou stain, 200x)

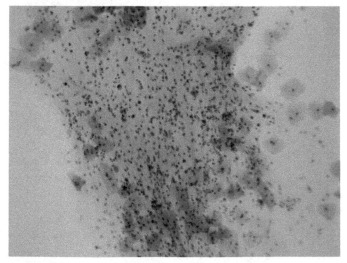

Fig. 6. Background mucin streaks are very thick and have characteristic color and distribution. (Papanicolaou stain, 100x)

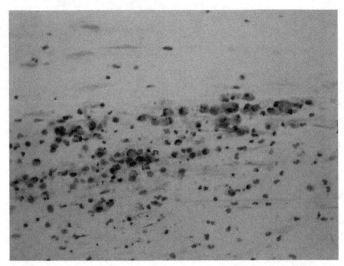

Fig. 7. Ileal conduit urine like features contained abundant degenerative glandular mucous cells and debris. (Papanicolaou stain, 200x)

Recognizing this dirty mucin background would be very important in the interpretation of AGC Pap smears and helped the clinicians successfully found the primary site of cancers.

Endometrial debris admixed with atypical endometrial glandular cells would be seen not only in the cancer patients but also in benign lesions, such as endometrial polyp and intra-uterine contraceptive device (Figure 9-10). Clinical information is very important to avoid over diagnosis.

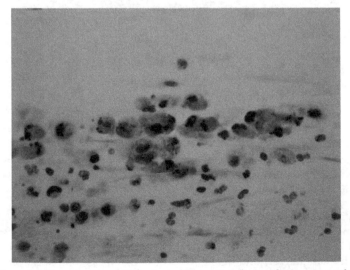

Fig. 8. These degenerative mucous cells had small eccentric hyperchromatic nuclei and abundant mucous cytoplasm indicating mucinous glandular origin. (Papanicolaou stain, 400x)

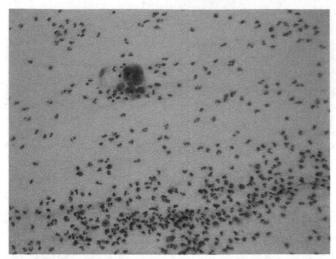

Fig. 9. Degenerative atypical endometrial glandular cells and endometrial debris were noted in a smear of patient with intra-uterine contraceptive device (IUD). Originally, the diagnosis of AGC-NOS, EM was given without knowing the IUD situation. (Papanicolaou stain, 200x)

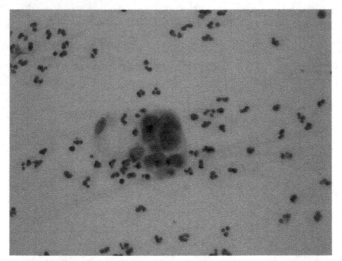

Fig. 10. Higher magnification showed small three-dimensional group of endometrial glandular cells with characteristic cytoplasmic vacuolation and slightly enlarged hyperchromatic nuclei. (Papanicolaou stain, 400x)

The major differential diagnoses of atypical endocervical glandular cells include adenocarcinoma in situ (Figure 11), tubal metaplasia (Figure 12) and lower uterine segment cells (Figure 13). When the lesion is adequately sampled and the abnormal cells are well visualized both quantitatively and qualitatively, the diagnosis will be no problem in most circumstances. Otherwise, these look-alike entities should be taken into the list of differential diagnoses.

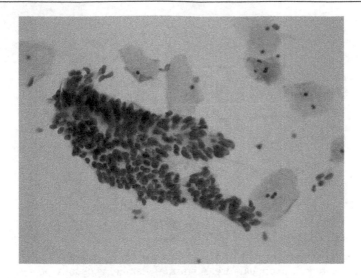

Fig. 11. Super-crowded endocervical glandular cells presenting pseudostratification and feathering edge. The nuclei are elongated and hyperchromatic with high N/C ratio.

Nucleoli are absent. However, only two fragments were seen. The diagnosis of AGC-FN was made. Final histology proved to be endocervical adenocarcinoma in situ. (Papanicolaou stain, 200x)

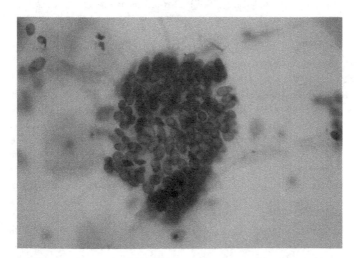

Fig. 12. Tubal metaplasia can closely mimic adenocarcinoma in situ. However, on close inspection, the abnormalities, such as crowding, nuclear elongation, hyperchromasia, and stratification are less severe. Locating terminal bars or cilia can help in confirmation, but they could not be identified in the smear we examined. The original diagnosis for this case was AGC-NOS, endocervical origin. (Papanicolaou stain, 400x)

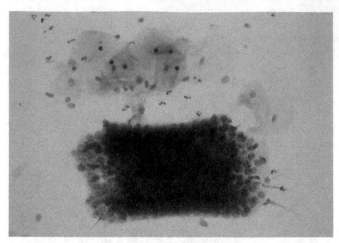

Fig. 13. Lower uterine segment cells were composed of tightly packed uniform glandular cells in crowded honeycomb appearance. A stromal element is usually present in the surrounding area, which is an aid in the differential diagnosis. However, the stromal component is absent in the present case, the original diagnosis was AGC-NOS, cell origin also not otherwise specified. (Papanicolaou stain, 400x)

4. Discussion

The TBS was first introduced in 1988 for reporting cervical/vaginal cytology findings (National Cancer Institute Workshop, 1989). Revisions were made in 2001 to improve its sensitivity and specificity. In terms of categories of atypical glandular cells, In the 1988 version these cells were defined as "atypical glandular cells of undetermined significance (AGUS)". In the current 2001 version (Solomon, 2002), this nomenclature has changed to "atypical glandular cells, not otherwise specified (AGC-NOS)" and "atypical glandular cells, favor neoplastic (AGC-FN). Subclassification of cell origin (endocervical, endometrial, or not otherwise specified) should be done whenever possible. The TBS 2001 reporting system was proved to be better for detecting underlying gynecological lesions, including precursors and invasive malignant diseases in many reports (Behtash, 2007; Duska, 1998; Jeng, 2003; Koonings, 2001; Lai, 2008; Manetta, 1999; Mood, 2006; Soofer, 2000). The incidence of abnormal pathology ranged from 8.2% to 53%. Most studies defined abnormal pathology as precursors and invasive malignant diseases. Low grade squamous intra-epithelial lesions (LSIL) were excluded. The invasive malignancies may originate not only from uterine cervix and corpus but also extra-uterine organs, such as fallopian tube, ovary, colon and rectum. In this current report, we found that endometrial cancer was by far the most common malignant disease, 38 in 53 cases (72%), diagnosed in the AGC smears (Table 2). This was the highest data ever reported. The reason may be that since our previous observation of the importance of endometrial debris (Lai, 2008), the screeners paid much attention to this kind of cytology findings. The sensitivity of reporting endometrial debris increased and then detected more endometrial cancers.

The concept of subclassification of AGC to NOS and FN categories as an important predictor for the risk of abnormal pathology was further supported in the current study (Table 3). The subgroup of AGC-N, EM had the highest rate of abnormal pathology, followed by AGC-NOS, EM, AGC-N and AGC-NOS; 18 of 21 (86%), 30 of 75 (40%), 4 of 11 (36%) and 24 of 83 (29%), respectively. The difference was significant. (P<0.001) Women with AGC-N were more likely to have significant pathology (22 in 32 (69%)) compared with those with AGC-NOS (54 in 158 (34%)). It was statistically significant. (P<0.001) The results were in accordance with other previous studies (Adhya, 2009; Behtash, 2007; Sawangsang, 2011; Westin, 2008; Zhao, 2009). They also confirmed the high risk nature of AGC smears. They have consistently demonstrated that the rate of abnormal pathology was significantly high if the AGC smears further classified as favoring neoplasia (41%-70%).

Since we have observed the importance of endometrial debris in our previous report (Lai, 2008), in the current study, only endometrial origin smears were calculated separately in order to strengthen the importance of this factor. The endocervical and not otherwise specified origins were counted together. The subgroup of AGC-N, EM had the highest rate of abnormal pathology, followed by AGC-NOS, EM, AGC-N and AGC-NOS; 18 of 21 (86%), 30 of 75 (40%), 4 of 11 (36%) and 24 of 83 (29%), respectively. The difference was significant. (P<0.001) The endometrial origin addressed cases had more abnormal pathology results than those not being addressed, 48 of 96 (50%) v.s. 28 of 94 (30%). (P=0.004) (Table 3) The results further confirmed the importance of cell origin of AGC, especially the endometrial origin. The predictive value of background endometrial debris such as histiocytes for endometrial pathology in Pap smears has been a subject of controversy. Some studies have suggested a significant finding but the others didn't (Iavazzo, 2008; Koss, 1962; Nassar, 2003; Ng, 1974; Nguyen, 1998; Wen, 2003). The controversy is understandable, because the biopsy rate and the yield of endometrial neoplasm in these patients were relatively low in the past. According to Browne's study (Browne, 2005), they found a 5-fold increase in the frequency with which endometrial cells were reported after the implementation of the TBS 2001. This then resulted in 25.2% of biopsies, a 1.3 fold increase in the overall number of tissue proof. Our another study (Lai, in press) of the importance of endometrial debris also revealed similar results. It showed that even in the absence of AGC, the presence of endometrial debris rather than the menopausal status was more related to the rate of biopsy procedure and a malignant pathology result. Degenerative necrotic debris is a significant risk factor for endometrial pathology, regardless of the presence or absence of AGC. Although the cervical screening program is not designed to detect endometrial lesions, early detection of any such cases is possible and is a bonus to be beneficial for these patients by identifying the significant degenerative endometrial debris in the Pap smears. Finally, another characteristic dirty mucinous background cytomorphology indicating endocervical adenocarcinoma observed in our series, being limited number of cases at present, will be continuously studied to investigate its sensitivity and specificity.

5. Conclusion

In the current study, AGC smears were associated with a high prevalence of abnormal pathology, including precursors and malignant diseases. Furthermore, the results of separations between "not otherwise specified" v.s. "favor neoplastic' and "endometrial origin" was statistically significant. Endometrial cancers were the most common neoplasm in the AGC patients. The above findings supported the TBS 2001 for subcategories of AGC and subclassifications of cell origins and the appropriate management algorithm of the guidelines of ASCCP 2006.

6. References

Adhya, A.K., Mahesha, V., Srinivasan, R., Nijhawan, R., Rajwanshi, A., Suri, V., & Dhaliwal, L.K. (2009). Atypical glandular cells in cervical smears: histological correlation and a suggested plan of management based on age of the patient in a low-resource setting, *Cytopathology*, Vol. 20, No. 6, (December 2009), pp. 375-379, ISSN 1365-2303

Behtash, N., Nazari, Z., Fakhrejahani, F., Khafaf, A., & Azar, E.G. (2007). Clinical and histological significance of atypical glandular cell on Pap smear, *Australian and New Zealand Journal of Obstetrics and Gynaecology*, Vol. 47, No. 1, (February 2007), pp. 46-49, ISSN 0004-8666

Browne, T.J., Genest, D.R., & Cibas, E.S. (2005). The clinical significance of benign-appearing endometrial cells on a Papanicolaou test in women 40 years or older, *American Journal of Clinical Pathology*, Vol. 124, No. 6, (January 2006), pp. 834-837, ISSN 0002-9173

Duska, L.R., Flynn, C.F., Chen, A., Whall-Strojwas, D., & Goodman, A. (1998). Clinical evaluation of atypical glandular cells of undetermined significance on cervical cytology, *Obstetrics and Gynecology*, Vol. 91, No. 2, (February 1998), pp. 278-282, ISSN 0029-7844

Iavazzo, C., Kalmantis, K., Ntziora, F., Balakitsas, N., & Paschalinopoulos, D. (2008). Detection of large histiocytes in pap smears: role in the prediction of endometrial pathology?, *Bratislavske Lekarske Listy*, Vol. 109, No. 11, (February 2009), pp. 497-498, ISSN 0006-9248

Jeng, C.J., Liang, H.S., Wang, T.Y., Shen, J., Yang, Y.C., & Tzeng, C.R. (2003). Cytologic and histologic review of atypical glandular cells (AGC) detected during cervical cytology screening, *Int J Gynecol Cancer*, Vol. 13, No. 4, (August 2003), pp. 518-521, ISSN 1048-891X

Koonings, P.P., & Price, J.H. (2001). Evaluation of atypical glandular cells of undetermined significance: is age important?, *American Journal of Obstetrics and Gynecology*, Vol. 184, No. 7, (June 2001), pp. 1457-1459; discussion 1459-1461, ISSN 0002-9378

Koss, L.G. (1989). The Papanicolaou test for cervical cancer detection. A triumph and a tragedy, *JAMA*, Vol. 261, No. 5, (February 1989), pp. 737-743, ISSN 0098-7484

Koss, L.G., & Durfee, G.R. (1962). Cytologic diagnosis of endometrial carcinoma. Result of ten years of experience, *Acta Cytologica*, Vol. 6, No., (November 1962), pp. 519-531, ISSN 0001-5547

Lai, C.R., Hsu, C.Y., Tsay, S.H., & Li, A.F. (2008). Clinical significance of atypical glandular cells by the 2001 Bethesda System in cytohistologic correlation, *Acta Cytologica*, Vol. 52, No. 5, (October 2008), pp. 563-567, ISSN 0001-5547

Lai, C.R., Hsu, C.Y., & Li, A.F. (in press). Degenerative necrotic endometrial debris in Pap smear: The role in the prediction of endometrial pathology, American Journal of Clinical Pathology , (in press), ISSN: 0002-9173

Manetta, A., Keefe, K., Lin, F., Ahdoot, D., & Kaleb, V. (1999). Atypical glandular cells of undetermined significance in cervical cytologic findings, *American Journal of Obstetrics and Gynecology*, Vol. 180, No. 4, (April 1999), pp. 883-888, ISSN 0002-9378

Mood, N.I., Eftekhar, Z., Haratian, A., Saeedi, L., Rahimi-Moghaddam, P., & Yarandi, F. (2006). A cytohistologic study of atypical glandular cells detected in cervical smears during cervical screening tests in Iran, *Int J Gynecol Cancer*, Vol. 16, No. 1, (February 2006), pp. 257-261, ISSN 1048-891X

Nassar, A., Fleisher, S.R., & Nasuti, J.F. (2003). Value of histiocyte detection in Pap smears for predicting endometrial pathology. An institutional experience, *Acta Cytologica*, Vol. 47, No. 5, (October 2003), pp. 762-767, ISSN 0001-5547

National Cancer Institute Workshop (1989). The 1988 Bethesda System for reporting cervical/vaginal cytological diagnoses. , *JAMA*, Vol. 262, No. 7, (August 1989), pp. 931-934, ISSN 0098-7484

Ng, A.B., Reagan, J.W., Hawliczek, S., & Wentz, B.W. (1974). Significance of endometrial cells in the detection of endometrial carcinoma and its precursors, *Acta Cytologica*, Vol. 18, No. 5, (September 1974), pp. 356-361, ISSN 0001-5547

Nguyen, T.N., Bourdeau, J.L., Ferenczy, A., & Franco, E.L. (1998). Clinical significance of histiocytes in the detection of endometrial adenocarcinoma and hyperplasia, *Diagnostic Cytopathology*, Vol. 19, No. 2, (August 1998), pp. 89-93, ISSN 8755-1039

Sawangsang, P., Sae-Teng, C., Suprasert, P., Srisomboon, J., Khunamornpong, S., & Kietpeerakool, C. (2011). Clinical significance of atypical glandular cells on Pap smears: Experience from a region with a high incidence of cervical cancer, *Journal of Obstetrics and Gynaecology Research*, Vol. 37, No. 6, (June 2011), pp. 496-500, ISSN 1341-8076

Solomon, D., Davey, D., Kurman, R., Moriarty, A., O'Connor, D., Prey, M., Raab, S., Sherman, M., Wilbur, D., Wright, T., Jr., & Young, N. (2002). The 2001 Bethesda System: terminology for reporting results of cervical cytology, *JAMA*, Vol. 287, No. 16, (April 2002), pp. 2114-2119, ISSN 0098-7484

Soofer, S.B., & Sidawy, M.K. (2000). Atypical glandular cells of undetermined significance: clinically significant lesions and means of patient follow-up, *Cancer*, Vol. 90, No. 4, (August 2000), pp. 207-214, ISSN 0008-543X

Wen, P., Abramovich, C.M., Wang, N., Knop, N., Mansbacher, S., & Abdul-Karim, F.W. (2003). Significance of histiocytes on otherwise-normal cervical smears from postmenopausal women. A retrospective study of 108 cases, *Acta Cytologica*, Vol. 47, No. 2, (April 2003), pp. 135-140, ISSN 0001-5547

Westin, M.C., Derchain, S.F., Rabelo-Santos, S.H., Angelo-Andrade, L.A., Sarian, L.O., Oliveira, E., & Zeferino, L.C. (2008). Atypical glandular cells and adenocarcinoma in situ according to the Bethesda 2001 classification: cytohistological correlation and clinical implications, *European Journal of Obstetrics, Gynecology, and Reproductive Biology*, Vol. 139, No. 1, (July 2007), pp. 79-85, ISSN 0301-2115

Wingo, P.A., Cardinez, C.J., Landis, S.H., Greenlee, R.T., Ries, L.A., Anderson, R.N., & Thun, M.J. (2003). Long-term trends in cancer mortality in the United States, 1930-1998, *Cancer*, Vol. 97, No. 12 Suppl, (June 2003), pp. 3133-3275, ISSN 0008-543X

Wright, T.C., Jr., Massad, L.S., Dunton, C.J., Spitzer, M., Wilkinson, E.J., & Solomon, D. (2007). 2006 consensus guidelines for the management of women with abnormal cervical screening tests, *J Low Genit Tract Dis*, Vol. 11, No. 4, (October 2007), pp. 201-222, ISSN 1089-2591

Zhao, C., Florea, A., Onisko, A., & Austin, R.M. (2009). Histologic follow-up results in 662
    patients with Pap test findings of atypical glandular cells: results from a large
    academic womens hospital laboratory employing sensitive screening methods,
    Gynecologic Oncology, Vol. 114, No. 3, (September 2009), pp. 383-389, ISSN 1095-6859

7

Cytology of Cervical Intraepithelial Glandular Lesions

Ana Ovanin-Rakić
Department of Gynecologic Cytology,
Department of Gynecologic and Perinatal Pathology, Zagreb,
Croatia

1. Introduction

Cytology of the cervix has been properly recognized and accepted in the detection and follow-up of squamous intraepithelial lesions, whereas its role in endocervical cylindrical epithelium is less well defined. Glandular lesions of the cervix uteri have been showing a rising incidence for the last 20 years, especially among young women (Nieminen et al.1995). This trend could be attributed to several factors: better diagnosis with more appropriate techniques of sampling and specimen preparation for both cytological and histological analysis, better recognition of precursor lesions, changes in nomenclature, evolving methods of treatment and an improved understanding of morphological features, having all led to the development of criteria for the diagnosis of early dysplastic lesions. Another reason for the observed rise is the increased prevalence of these lesions.

As the major purpose of the Papanicolaou smear tests is the earliest possible diagnosis of cervical cancer and its precursors, both cervix and endocervix must be adequately sampled as the most common sites of these lesions. The best time for obtaining a smear is midcycle, i.e., two weeks after the first day of the last menses. Ideally, the woman should not have had intercourse, used douches, vaginal medication, or vaginal contraceptives 48 hours prior to obtaining a smear. It is vital that detailed clinical information be provided to the cytology laboratory. This information should include: date of the last menstrual period, results of previous Papanicolaou smears, history of fertility treatments or hormone therapy, history of abnormal bleeding, usage of intrauterine contraceptive devices, history of malignancy of female genitalia, of hysterectomy, radiation, and the results of any previous cervical biopsy.

The accuracy of clinical cytology relies to a large extent on sucessful sampling in obtaining the Papanicolaou smear and on its proper fixation and staining. A specimen from the cervicovaginal area that has been satisfactorily obtained and prepared for microscopic examination exhibits an abundance of well-preserved and meticulously stained diagnostic cellular material that remains preserved for indefinite slide storage and later review.

Glandular lesions are frequently detected in histology of cytologically diagnosed squamous intraepithelial lesions (SIL).

Cytological criteria for the identification of glandular intraepithelial lesions (GIL) have not yet been fully articulated, especially for the precursors of adenocarcinoma in situ (AIS), and these lesions may frequently remain unrecognized. Documenting the sequence of neoplastic events in the endocervix poses problems, except for its lowest segment, because the endocervical canal cannot be visualized by colposcopy and, therefore, cytological sampling cannot be targeted. Also, in spite of numerous efforts, morphological recognition of sequential abnormalities of endocervical cells is much more difficult than in squamous cells (Lee, 1999). Primary factors that contributed to either screening errors or diagnostic errors in AIS were insufficient quantities of material or poorly preserved abnormal material and aggregates of glandular cells. (Ruba et al., 2004).

2. Classification

By analogy to squamous cell cervical cancer precursors that demonstrate a wide spectrum of histological changes, some authors have proposed parallel classification schemas for endocervical adenocarcinoma precursors that include lesions with a lesser degree of abnormality than AIS. Such low grade putative glandular precursor lesions were termed endocervical dysplasia (Bousfield et al., 1980), cervical intraepithelial glandular neoplasia - CIGN (Gloor & Hurlimann, 1986), endocervical columnar cell intraepithelial neoplasia - ECCTN (van Aspert - van Erp et al., 1995), low grade glandular intraepithelial lesion - LGIL (DiTomasso et al., 1996), endocervical glandular dysplasia - EGD (Casper et al., 1997), endocervical glandular atypia - EGA (Goldstin et al., 1998), and cervical glandular intraepithelial neoplasia Low grade - L-CGIN (Kurian & al-Nafussi, 1999). We prefer the term glandular intraepithelial lesion (GIL) grade 1 and 2.

In contrast to squamous intraepithelial lesions with identifiable subgroups, in the case of glandular epithelium only adenocarcinoma in situ, included in the NCI Bethesda 2001 cytological classification, has been recognized. (http://bethesda 2001 .cancer.gov)

A uniform classification of cervical cytology findings known as Zagreb 1990 (Audy-Jurkovic et al., 1992) and developed by combining the original 1988 Bethesda System (TBS) classification (NCI, 1989) and our previous classification (Audy-Jurkovic et al., 1986) has been used in Croatia since 1990. As the TBS has been supplemented and/or modified on several occasions since its introduction (NCI, 1993 2001; Kurman & Solomon, 1994), we considered it plausible to revise our classification accordingly, i.e. by modifying and/or supplementing points of dispute noted over the past years, and by harmonizing it with the NCI Bethesda System 2001.

The current classification, **Zagreb 2002**, has been introduced as a new uniform classification system of cervical cytology findings used in Croatia (Fig. 1) (Ovanin-Rakic et al., 2003).

General classification consists of two groups, **"negative"**, for intraepithelial or invasive lesions, and **"abnormal cells"**, the latter referring to all cell alterations that are morphologically consistent with intraepithelial or invasive malignant lesions. The "negative" group refers to findings which are within normal limits, cell alterations associated with particular reactive and reparatory reactions, and with the presence of cells indicative of certain risks (e.g., findings of endometrial cells of benign appearance beyond the cycle or in the postmenopausal period).

COMPLETED BY GYNECOLOGIST

| Name _____ Date of birth _____ City_____ |
| Address _____ Tel./Fax/e-mail_____ Date _____ |
| Health unit _____ Patient's number _____ |

Parity	Menstrual cycle	First day of LMP	Postmenopausis	Sample	Identification number	Laboratory number

Contraceptives: ☐ hormones ☐ IUD ☐ other ☐ without   ☐ V _____ _____
Previous diagnostic-therapeutic procedures   ☐ C _____ _____
Cytological diagnosis_____   ☐ E _____ _____
Histological diagnosis_____   ☐ Vulva _____ _____
Other _____   ☐
Therapy _____   Clinical diagnosis
Supravital analysis   ☐ Within normal limits
L 1 2 3   ☐ Colposcopy   ☐ Other
☐ Gardnerella vag.   ☐ Endocervicoscopy   Clinical remarks:
☐ Trichomonas vag.
☐ Fungi
☐......................   Date: Gynecologist's Signature

SPECIMEN ADEQUACY

☐ **Satisfactory for interpretation**
☐ **Unsatisfactory for interpretation**
  ☐ Specimen rejected/not processed
  ☐ Specimen examined, but evaluation of epithelial abnormality is not possible
  Comments for the specimen adequacy:
    ☐ Incorrect label
    ☐ Broken slide
    ☐ Poorly fixation or inadequately preserved
    ☐ Scant cellularity
    ☐ No endocervical cells
    ☐ Obscuring inflammation
    ☐ Obscuring blood
    ☐ Thick areas
    ☐ Presence of foreign material
    ☐ Other:_____

GENERAL CATEGORIZATION

☐ **Negative for Intraepithelial Lesion or Malignancy**
☐ **Abnormal cells (See descriptive diagnosis)**

DESCRIPTIVE DIAGNOSIS

MICROORGANISMS

☐ Bacillus vaginalis   ☐ Gardnerella vaginalis
☐ Mixed flora   ☐ Chlamydia trachomatis
☐ Fungi   ☐ Cellular changes associated with HSV
☐ Trichomonas   ☐ Cellular changes associated with HPV
☐ Actinomyces   ☐ Other:_____

Other non-neoplastic changes
☐ **Reactive cellular changes associated with:**
  ☐ Inflammation   ☐ IUD
  ☐ Radiation   ☐ Other: _____

☐ **Reparation**   ☐ **Reserve cells**
☐ **Parakeratosis** ☐ **Diskeratosis** ☐ **Hyperkeratosis**
☐ **Glandular cells post hysterectomy**
☐ **Endometrial cells**
  ☐ out of phase in menstrual patient   ☐ in Postmenopausis
☐ **Cytohormonal status incompatible with age & anamnesis**
☐ **Other**

Abnormal cells

☐ **Squamous cells**
  ☐ Atypical squamous cells (ASC)
    ☐ Of undetermined significance (ASC-US)
    ☐ Cannot exclude HSIL (ASC-H)
    ☐ Cannot exclude invasion
  ☐ Squamous intraepithelial lesion (SIL)
    ☐ Dysplasia levis → CIN I → ☐ low-grade SIL
    ☐ Dysplasia media → CIN II
    ☐ Dysplasia gravis
    ☐ Carcinoma in situ } CIN III > ☐ high-grade SIL
    ☐ Plus: cellular changes associated with HPV
    ☐ Cannot exclude early invasion
  ☐ Carcinoma planocellulare

☐ **Glandular cells**   Origin:
☐ Atypical glandular cells   ☐ Endocervical
  ☐ Favor reactive
  ☐ Favor intraepithelial lesion   ☐ Endometrial
  ☐ Favor invasive lesion   ☐ Ekstrauterine
☐ Adenocarcinoma in situ (AIS)   ☐ Not otherwise specified
☐ Adenocarcinoma
☐ **Atypical cells of undetermined significance** _____
☐ **Other malignant neoplasm** _____

Sample boxes: V / C / E

RECOMONDATION

☐ Repeat Smear   ☐ Colposcopy
☐ Repeat after therapy   ☐ Histology
☐ Repeat in 4 months   ☐ Further examination
☐ Repeat in 6 months   ☐ Other
☐ Regular control

RECOMONDATION:

| Admitted: | Replied: | Cytotechnologist's Signature: | Cytologist's Signature: |

Fig. 1. Uniform classification of cytological findings of cervix uteri "Zagreb 2002", modification of the "Zagreb 1990" and "NCI Bethesda System 2001"

Unlike NCI 2001, we have kept the term **"diagnosis"** instead of "interpretation/finding result". Descriptive diagnosis contains the subgroups of **"microorganisms"** (microorganisms that can be identified directly or on the basis of a specific cytopathic effect); **"other non-neoplastic findings"** (reactive cell alterations, reparatory epithelium, reserve cells, parakeratosis, dyskeratosis, hyperkeratosis, post-hysterectomy cylindrical cells, endometrial cells beyond the cycle or in the postmenopausal period, and cytohormonal status inconsistent with age and/or history), and **"abnormal cells"** (squamous, glandular, abnormal cells of undetermined significance, and other malignant neoplasms).

In the **Zagreb 2002** classification, like in the NCI 2001, glandular lesions have been divided into three categories: **"atypical glandular cells"** (AGC), **"adenocarcinoma *in situ*"** (AIS), and **"adenocarcinoma"**. In the case of squamous epithelium, and unlike in NCI 2001, AGC have been divided into three subgroups, instead of two:

- favor reactive – cell alterations that are more pronounced than benign reactive ones but quantitatively and qualitatively less pronounced than those in intraepithelial lesions;
- favor intraepithelial (GIL1,GIL2) – cell alterations of low to moderate severity, without inflammatory cell changes, and/or suggestive of AIS, without definite criteria;
- favor invasive – cell alterations suggestive of invasive lesions, where differential cytological diagnosis cannot be made, mostly due to poor specimen preparation.

The group of adenocarcinoma in situ requires the establishment of well defined criteria.

The group of adenocarcinoma invasivum has not been modified relative to previous classifications.

For any group or subgroup of abnormal glandular cells, it is crucial to identify the origin of cylindrical epithelium whenever possible, as it is of great importance for further diagnostic and therapeutic procedures. At the end of the report, the cytologist provides the clinician with instructions on how to improve the quality of cervicovaginal smears, with guidelines on further procedures for a particular cytological finding. These instructions are in line with the current diagnostic-therapeutic protocols in use in Croatia (Ljubojevic et al., 2001).

Assessment of specimen adequacy is one of the substantial qualitative components of a finding. All criteria advocated by NCI Bethesda System 2001 (NCI, 1989,1993, 2001; Solomon et al., 2002) have been incorporated into our classification system. Information on the components of the transformation zone, i.e. finding of endocervical cylindrical epithelial cells, improves overall specimen quality thereby stimulating efforts to obtain an optimal specimen. However, the absence of such information is by no means a reason for a repeat smear (Pajtler & Audy-Jurkovic, 2002).

Cytodiagnosis of cervical cylindrical epithelial lesions lags behind the cytodiagnosis of squamous epithelial lesions both in terms of screening and differential diagnosis. The Australian (Roberts et al., 2000) modification of TBS (NCI, 1989) for glandular lesions points to the risks in the presence of high-grade abnormalities, thus resulting in more appropriate recommendations and protocols.

Cytological diagnosis of adenocarcinoma in situ of endocervical cylindrical epithelium as a separate entity was only included in the NCI Bethesda System 2001 classification, whereas dysplasia of endocervical cylindrical epithelium as an AIS precursor is still considered cytologically and histologically to be an inadequately defined entity (Zanino, 2000) and has not been included in the classification (NCI, 2001).

In most cases, morphological characteristics allow for differentiation between atypical endometrial cells and endocervical cells (Chieng & Cangiarella, 2003).

The proposed **Zagreb** classification, with amended and/or supplemented points of dispute identified in previous classifications, is uniform for Croatia. It allows for both internal and external performance quality control, along with appropriate reproducibility of cervical cytology relative to terminology adopted worldwide.

3. Epidemiology

The prevalence of AIS is not known, but is considerably lower than the prevalence of SIL. In the Surveillance Epidemiology End Results (SEER) public database, which contains data from patients entered into the database between 1973 and 1995 (Plaxe & Saltzstein, 1999), the ratio of in situ and invasive lesions is 1:3 for glandular and 5.25:1 for squamous lesions.

The rate of dysplasia of endocervical cylindrical epithelium is 16-fold that of AIS. and the mean age at diagnosis for endocervical glandular dysplasia is 37 (Brown & Wells, 1986).

The mean age at diagnosis of women with AIS in the SEER registry is 38.8, , and it is 51.7 for invasive adenocarcinoma (AI) of the cervix.

The median age of patients in our study (Ovanin-Rakic et al., 2010) was 40, which is comparable to 41, reported in the literature (Kurian & al-Nafussi, 1999), and was slightly higher than the averages from other studies (Shin et al., 2002).

Patients diagnosed with mild glandular lesions (GIL1) are on the average 10 years younger than those with the invasive disease. The mean age of AIS patients is about 13 years younger than in those with AI of the cervix. The age differnce between AIS and AI patients suggests the former to be a precursor lesion. It takes about 13 years for the AIS as an adenocarcinoma precursor to progress to AI. Such a long period of carcinogenesis recorded for lesions of endocervical cylindrical epithelium provides opportunities for their early detection and results in the reduction of incidence of AI. Additional support for implicating AIS as precursor of AI comes from several reports which had cytological or histological evidence of AIS appearing 2 - 8 years before detection of the invasive lesions (Boon et al., 1981).

Epidemiological risk factors for cervical adenocarcinoma include those that correlate with the risk of acquiring Human Papillomavirus (HPV) infections, such as multiple sexual partners and engaging in sexual intercourse at an early age. In addition, adenocarcinoma was also found to be associated with obesity and with the prolonged use of oral contraceptives.

Recent trials evaluating the efficacy of virus-like particle vaccines in the prevention of persistent infection with HPV-16 and HPV-18 in young women have been shown to be highly effective.

4. Etiology

In a series of initial cervical swabs, minimal to severe atypias of cylindrical epithelium were detected in 50% of cases with squamous epithelial lesions (Pacey & Ng, 1997), pointing to common etiological factors.

The incidence of coexisting squamous lesions was 74.8% in our study (Ovanin-Rakic, 2010), comparable to the 41 – 76.7% reported in the literature (Im et al.,1995; Shin et al., 2002).

The etiology of squamous cell carcinoma of the cervix, the most common type of cervical malignancy, is linked to infection with oncogenic types of HPV, but the pathogenesis of adenocarcinoma is less well understood. Pirog et al., 2000, detected a very high prevalence of HPV DNA in cervical adenocarcinoma relative to most previous reports. The relative difficulty in detecting HPV DNA in adenocarcinoma, in contrast to squamous cell carcinomas, may be attributed to lower viral load in glandular lesions as compared to squamous lesions. Premalignant and malignant squamous lesions, in particular those associated with HPV 16, contain a large number of episomal viral particles, in addition to integrated HPV sequences (Stoler et al., 1992). Glandular epithelium does not support productive viral infection, and HPV DNA in endocervical neoplasms (notably HPV 18) is usually present in integrated form (Park et al., 1997).

Associations between endocervical glandular atypia (dysplasia) and HPV are more contraversial. In the original study by Tase et al., 1989, only 2 of 36 cases of endocervical dysplasia contained HPV DNA. However, another study (Higgins et al., 1992) reported that 94% were associated with HPV DNA and 75% were associated with HPV 18.

5. Clinical features and management

Most patients diagnosed with GIL are free from clinical symptoms, thus a lesion is detected by cytology on routine swab sampling ("PAP" smears), or by histology (endocervical curettage - ECC, biopsy specimen, conization specimen, loop excision, hysterectomy material) on examination for SIL, or during operative procedure for myoma. (Ovanin-Rakic et al., 2010) In women who are symptomatic, the most common complaint is abnormal vaginal bleeding, either postcoital, postmenopausal, or out of phase. In intraepithelial glandular lesions, the portion is of normal macroscopic appearance and colposcopic images have long been considered nonspecific. However, some authors state that characteristic vascular changes are found in glandular lesions (Singer & Monaghan, 2000). Cytology has a very prominent and responsible role in detection of these lesions.

The anatomical distribution of AIS showed that AIS involved both surface and gland epithelia, a variable number of quadrants, glands beneath the transformation zone in about two thirds of cases, was multifocal only occasionally, and extend up the endocervical canal for a variable distance up to 30mm (Bertrand et al., 1987; Im et al., 1995). Several reports suggest that women of childbearing age may safely be followed after cold-knife conisation with minimal risk provided that the margins are negative. The cone should be cylindrical, encompassing the entire transformation zone if possible, and the sampling depth of endocervical glands should be 5mm from the canal.

It should extend parallel to the endocervical canal for at least 25mm before a 90-degree turn toward the endocervical canal (Bertrand et al., 1987). If the diagnosis is established with a loop excision, even with negative margins, a cold-knife conisation should be performed. After the completion of childbearing, a hysterectomy is recommended because of the paucity of data concerning the long-term history of AIS. In those women for whom childbearing is not important, simple hysterectomy in the face of negative margins is acceptable (Östör et al., 2000; Shin et al., 2002). Some reports indicated that a deep surgical excision with negative margins might be sufficient treatment for some women. (Azodi et al., 1999).

The treatment of glandular leasons is more difficult than that of their squmous counterparts because of the younger age at diagnosis. Managemant of fertility is often an issue, with strong desire for conservation of the uterus. Careful documentation of discussions regarding the risk of conservative management is important as well as documentation of the need for hysterectomy once the childbearing is completed.

6. Cytological features

The interpretation of observed cells requires meticulous scientific training, dedication and experience. Reaching a definitive diagnosis utilizing cells that have desquamated freely from epithelial surfaces or cells that have been forcibly removed from various tissues, demands detailed examination of all available evidence. One of the most important aspects of cytological interpretation is the acquisition of comprehensive knowledge of the normal environment of the tissue to be examined. This knowledge has to take into account the diverse physiological as well as pathological settings that would normally be found in that particular tissue. Without such detailed understanding, the exercise of cytological interpretation can become a trap for a novice. In order to recognize the cytological appearance of endocervical glandular neoplasia with maximal sensitivity and specificity, a solid understanding of normal and variant normal morphology of the cells is necessary.

6.1 Normal columnar cells

The columnar epithelial cells characteristically have basally placed nuclei and tall, uniform, finely granular cytoplasm filled with mucinous droplets. The cells lining the luminal surface have been termed "picket cells" because of their resemblance to a picket fence. It is not known whether regeneration occurs from the underlying subcolumnar reserve cells.

The nuclei of endocervical cells are finely granular and of approximately the **same size as the nuclei of intermediate squamous cells**. The nuclei tend to form dense, dark, nipple-like protrusion that usually appears as a homogeneous extension of the nucleus into the adjacent cytoplasm. The remainder of the nucleus is usually less dense and has a normal appearance. (Boon ME & Gray W, 2003) .

6.1.1 Endocervical reserve cells

Rarely seen, endocervical reserve cells are young, endocervical, parabasal cells in close contact with the basement membrane.

They have multipotential differentiation and may be seen in sheets or in loose clusters of single cells. (Fig.6.1.1.1.) Their cytoplasm is adequate to scanty, cyanophilic and finely vacuolated. Their round to oval nuclei are centrally located, with fine, uniformly distributed chromatin. Small, round chromocenters are often multiple. Mitoses are occasionally seen and have no significance (Naib, 1996).

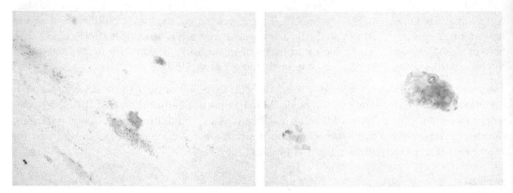

Fig. 6.1.1.1. Loose clusters of normal endocervical reserve sells. The mitotic figure has little diagnostic significance. (Papanicolaou x100, and x400).

6.1.2 Ciliated endocervical cells

Ciliated endocervical cells are the result of direct traumatic exfoliation. They can be single, in tight clusters, in small sheets, or in palisade formations when viewed from the side and honeycomb-like in appearance when their apical ends are in focus. Their size varies, but their shape is fairly constant, cylindrical or pyramidal. When a cell is well preserved, delicate pink cilia are attached to this lavender or red terminal bar or plate. (Fig.6.1.2.1.) This terminal bar can persist even after the cilia have been lost through degeneration. Their length varies according to the original position of the exfoliated cell in relation to the axes of the endocervical canal. The cytoplasm is elongated, with a semitransparent, lacy appearance and cytoplasmic borders that are thin and distinct, in contrast to those found in other types of endocervical cells. They stain darker than the pale mucus-producing endocervical cells. (Naib, 1996; Boon & Gray, 2003).

Fig. 6.1.2.1. Ciliated endocervical cells. Note the multincleation (Papanicolaou x100, x400).

Depending on the stage of maturation of the cell and its function, the nuclei are centrally placed or close to the apical cellular end, in contrast to the position of the nuclei in nonciliated cells. These nuclei are round or oval in shape and vary moderately in size. Their chromatin is finely granular and uniformly distributed. The nuclear borders are even and smooth, and they often merge with the cytoplasmic membrane on both sides. When multiple, the nuclei may overlap with little moulding.

Occasionally, nonsecretory cells with cilia are observed, the main function of which appears to relate to the distribution and mobilization of endocervical mucus. Rare, small, dark, nipple like protrusions may be seen in the nuclei of mature or reserve endocervical cells.

Detached ciliary tufts or ciliocytophthoria in cervicovaginal smears, are a very rare event and cannot be correlated with time of cycle or age of patient.(Fig.6.1.2.2.)

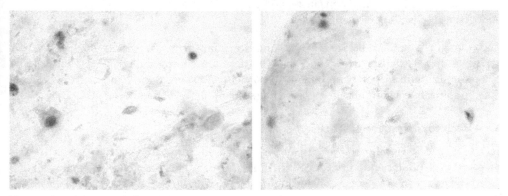

Fig. 6.1.2.2. Ciliocitophthoria. (Papanicolaou x1000).

6.1.3 Nonciliated endocervical cells

Nonciliated endocervical cells occur as single cells, in clusters, or in palisade formations and with a honeycomb-like appearance. (Fig.6.1.3.1.; Fig.6.1.3.2.)

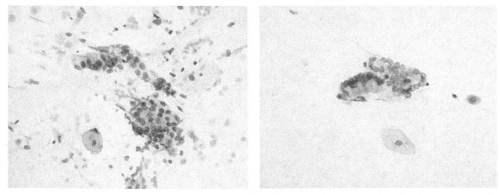

Fig. 6.1.3.1. Group of nonciciliated endocervical cells in sheet, palisade and rosettes. Note the same size nuclei of columnar and intermediate squamous cells. (Papanicolaou x400)

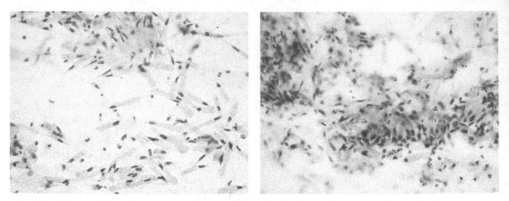

Fig. 6.1.3.2. Nonciliated singly, in cluster and palisade formation; very distended endocervical cells (Papanicolaou x400).

These long, columnar cells vary in size and are uniform in shape and elongated. Their adequate cytoplasm is narrow, and their borders are sharp, smooth, and delicate. The cytoplasm is semitransparent and finely vacuolated, and stains poorly and unevenly as pale blue. In some, fine acidophilic granules can be seen (Naib, 1996; Boon & Gray, 2003)

6.1.4 Secretory endocervical cells

Secretory endocervical cells are found in increased number with chronic irritation, pregnancy, glandular endocervical polyps, or intake of various hormones and contraceptive pills. They vary in size and exfoliate singly or in clusters. (Fig.6.1.4.1). Their shape varies from round to triangular. Their cytoplasm is usually distended by single or multiple small or large secretory vacuoles.

When degenerated, they may contain numerous, large, healthy polymorphonuclear cells. The borders of the cytoplasm are often indistinct, thin, and very delicate. Because of the fragility of the cytoplasm, it is common to find numerous stripped nuclei with only a wisp of transparent cytoplasm still attached in strands of thick cervical mucus in the smear.

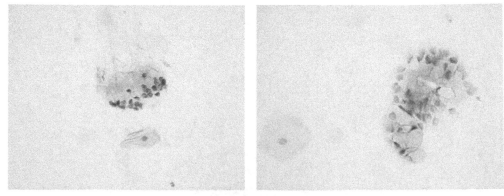

Fig. 6.1.4.1. Secretory endocervical cells in palisade and rosette formation.(Papanicolaou x400)

The nucleoli may be prominent, spherical in shape, and variable in number. Multinucleation is common, especially in cases of hormonal hyperplasia and chronic or acute cervicitis.

The nuclei are often enlarged, oval-to crescent-shaped, and eccentrically situated toward the narrow end of the cell as the result of the cell's displacement by the secretory vacuols.

The size of the nucleus may vary in diameter. The nuclear membrane is often fuzzy. The chromatin is coarsely clumped and has tendency to condense toward the nuclear membrane. Some of the nuclei may, in cases of hyper secretion, appear almost completely pyknotic with an extreme crescent-like shape. (Fig.6.1.4.2.)

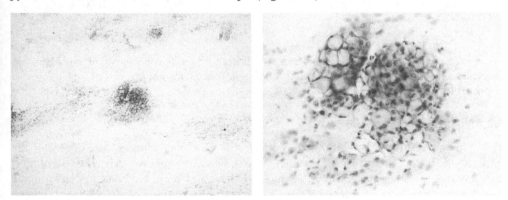

Fig. 6.1.4.2. Groups of secretory endocervical cells. Note the cytoplasmic secretory vacuoles. (Papanicolaou x100, and x400)

6.1.5 Endocervical stripped nuclei

Endocervical stripped nuclei, so-called bare, or naked, which often have a wisp of cytoplasm still attached, are commonly seen in smears from postmenopausal or pregnant women or from women with an endocervical ectropion.

These nuclei are uniformly round or oval but may vary moderately in size. Their nuclear membrane is regular and sharp with a small sign of degeneration. The chromatin pattern is uniform and finely granular with occasional clumping, similar to the normal nucleus of intact endocervical cell. (Fig.6.1.5.1.)

Condensation of the chromatin material toward the nuclear rim, pseudo hyperchromatism, or a clear, bland chromatic pattern may occur as a result of cellular degeneration. Occasional, single, small, central, reddish nucleoli can be seen in better-preserved naked nuclei.

These stripped nuclei should not be confused with poorly differentiated small-cell squamous carcinoma. Both may vary in size, but the shape of the endocervical nuclei is regular, with a smooth nuclear membrane, with chromatin finely granular, and uniformly distributed.

The cytological diagnosis of atypia should never be rendered from stripped nuclei alone.

An examination of better-preserved cells with intact cytoplasm is necessary. (Naib, 1996; Boon & Gray, 2003).

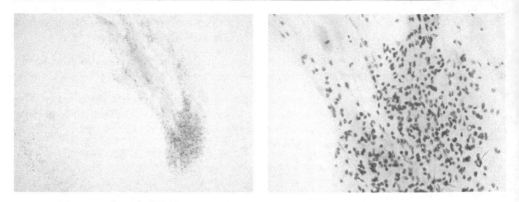

Fig. 6.1.5.1. "Stripped nuclei" of endocervical cells are uniformly round or oval. Note the chromatin pattern is uniform and finely granular similar to normal nucleus of intact endocervical cell. (Papanicolaou x 100, x 400)

6.2 Atypical Glandular Cells (AGC)

The cytologic features of atypical endocervical cells vary depending on the degree of the underlying histopathologic abnormality. The particular feature that may confound interpretation in these specimens is, again the presence of more crowded hyperchromatic groupings and the lack of spreading out that occurs in the making of conventional smears. This can lead to difficulty in identifying key nuclear and cytoplasmic features that could have otherwise made the interpretation more definitive, either toward benign/reactive or neoplastic. (Solomon, 2002; Chieng & Cangiarella, 2003; Waddell, 2003; Willson & Jones, 2004) .

6.2.1 Atypical Glandular Cells (AGC) favouring reactive process

These include endocervical cells from dense two- or three-dimensional aggregates, or sheets and palisades that have minor degrees of nucleolar overlapping. However, the changes may be reactive changes due to inflammation or trauma, as well as reflecting the earliest stages of GIL.(Fig.6.2.1.1.)

There is an increased number of intensely stained endocervical cells. Their abundant cytoplasm is dense, acidophilic, or overdistended by large secretory vacuoles, often containing well-preserved leukocytes or mucus secretions. Their nuclei are enlarged, with a smooth nuclear membrane and coarsely granular chromatin that is uniformly distributed and nucleolar feathering can be seen at the periphery of the cellular aggregates.

There is overlap between the nuclear features which may be seen in extreme inflammatory changes, and those which may be seen in some examples of glandular intraepithelial lesions.

(Fig.6.2.1.2.) They differ by regular distribution of their clumped chromatin and their smooth nuclear membrane. The nucleoli may be prominent, massive, spherical, and usually single, but they may vary in number.

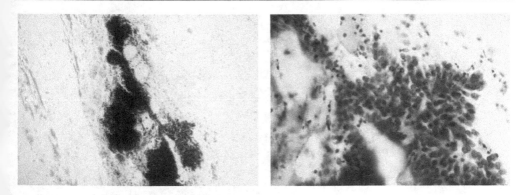

Fig. 6.2.1.1. Crowded sheets of endocervical cells. The nuclei are overlapping and hyperchromatic, but show only a mild variation in size within the sheet.

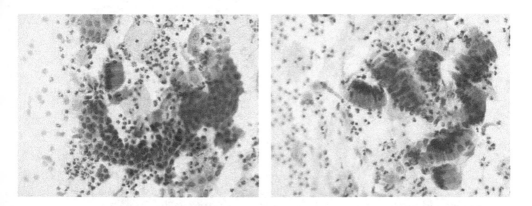

Fig. 6.2.1.2. In sheets and palisades pseudostratification of endocervical cells is present. Nuclei are slightly enlarged. (Papanicolaou x 100, x 400)

6.2.2 Atypical Glandular Cells (AGC), favouring intraepithelial lesions (glandular dysplasia)

Apearance of cytological atypias of the endocervical epithelium falls between those seen in normal glands and in AIS.

Mild (GIL1) and moderate (GIL2) glandular intraepithelial lesions have not been clearly defined, while reproducibility of the cytological and histological criteria for their identification has not been fully explored. Cellular alterations in GIL1 and GIL2 are similar to but less pronounced than those in AIS. The type of desquamation is also similar, except that the cylindrical cells are slightly packed showing a palisading pattern with mild pseudostratification, with less pronounced nuclear overlapping and observable feathering, rosettes, and glandular opening. (Fig.6.2.2.1.; Fig.6.2.2.2.; Fig.6.2.2.3.)

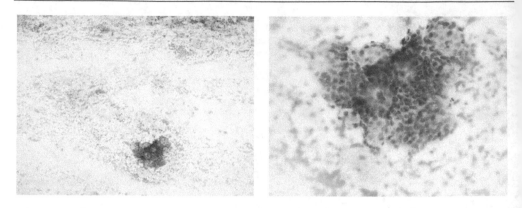

Fig. 6.2.2.1. A cluster of atypical endocervical cells (GIL 1) with glandular opening, with slight nuclear enlargement and overlapping. (Papanicolaou x100, x400)

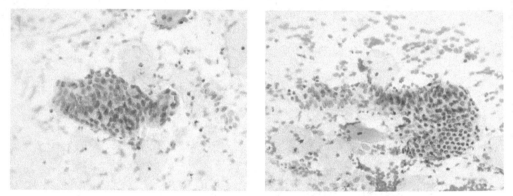

Fig. 6.2.2.2. Sheet of cells (left field) with slight nuclear enlargement and overlapping (GIL1) Note sheet of normal endocervical cells (right field) and palisade with slight nuclear enlargement and pseudostratification (GIL 1). (Papanicolaou x400)

According to the results some studies (Rabelo-Santos et al., 2008), feathering was the best criterion for predicting glandular neoplasia. Feathering was the criterion for distinguishing glandular from squamous neoplasia and also for distinguishing between glandular and . non-neoplastic diagnosis.

Rosettes and pseudostratified strips did not perfom as well. Some rosette formations can be seen in non-neoplastic cases. Squamous neoplasia, especially CIN 3 (cervical intraepithelial neoplasia), is frequently found to have rudimentary gland formation or micro-acinar structures, which can mimic AIS. These facts might help to explain the lower perfomance of the rosette when compared with feathering in the prediction of glandular neoplasia.

The cell size is like that in normal findings or slightly enlarged. **Nuclear size within a cluster varies to a greater extent than in the AIS** (Bousfield et al., 1980). The nucleus is round or oval, hyperchromasia is less pronounced, chromatin is finely granular and evenly distributed, and nucleoli are small and round. Mitoses are rare.

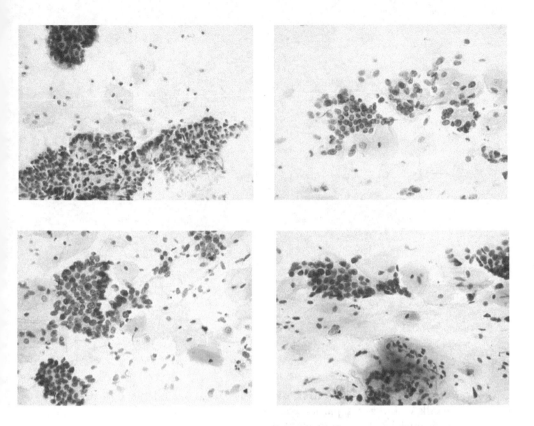

Fig. 6.2.2.3. The cervical smears contains sheets of crowded mild and moderate hyperchromatic endocervical cells with partial rosette and feathering.(GIL1, GIL2) A cluster of atypical cells (down right field) compared with normal cells in the same fields. (Papanicolaou x400).

Recognition of the characteristic architectural features in cell groups is very important in diagnosis. Without obvious and unequivocal nuclear change in endocervical cells, cytological diagnosis of GIL should not be made in the absence of these architectural features. Three-dimensional cell groups with disorderly cell arrangements, coarse grainy chromatin, and hyperchromasia with intercellular variation in nuclear staining intensity may be seen. None of the architectural abnormalities characteristic of GIL is present. . (Bousfield et al., 1980; Gloor & Hurlimann, 1986; vanAspert-van Erp et al., 1995; diTomassso et al., 1996; Golstein et al., 1998; Zaino, 2000).

Examples of abnormalities can usually be seen repeatedly in abnormal cellular material. This means that if the cellular material in question is scanty in a smear, a confident diagnosis of GIL may not be possible

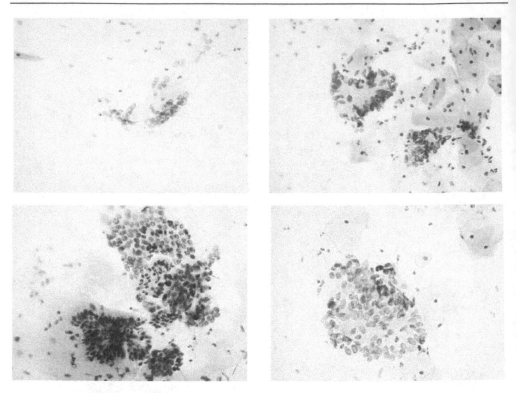

Fig. 6.2.6. The cervical smears (all four fields) contain sheets of moderate crowded endocervical cells with rosettes, partial rosettes and feathering with enlarged hyperchromatic nuclei. (GIL 2) A cluster of atypical cells compared with normal cells (pseudorosette and sheet) in the same fields (down left). Note the different nuclear size within a cluster (Papanicolaou x400).

6.2.3 Atypical Glandular Cells (AGC), favouring invasive lesions

The number of exfoliated diagnostic cells in smears varies according to the site, type, and size of the tumor and the technique used. Although generally larger than normal, the tumor cells, with few exceptions, imitate the appearance of the benign columnar cells from which they originate. In loose clusters or tight three-dimensional formations, abnormal degenerating columnar cells are identified in the "dirty" background of fresh and old degenerating blood cells, and cellular debris, consistent with the tumour diathesis.

The nuclei are enlarged, oval or round with eosinophilic granular cytoplasm. There is considerable nuclear overlapping and pleomorphism, and the chromatin pattern is coarsely granular, irregularly distributed and nucleoli are identifiable.

Unlike adenocarcinoma of the endometrium, the cells retain, especially at the periphery of clusters, their columnar configuration. A definitive cytological diagnosis cannot be made, mostly due to **poor specimen preparation.** (Fig.6.2.3.1.). Mitotic figures are occasionally seen.

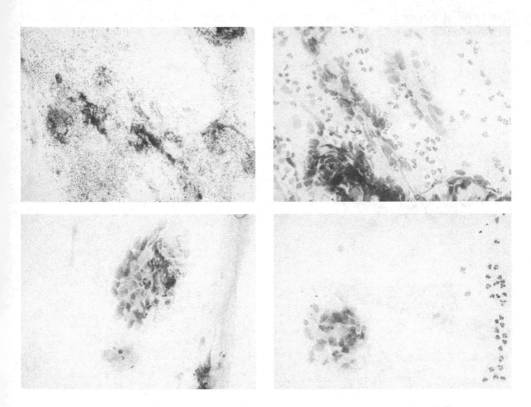

Fig. 6.2.3.1. Crowded and loose clusters, end rosettes of markedly atypical and degenerating endocervical cells. The nuclei are large, show irregular contours and coarse chromatin. Note the poor specimen preparation (Papanicolaou x100, x400 x400 x400).

6.3 Adenocarcinoma in situ

The cytomorphological criteria for diagnosis of AIS refers to changes in architectural features (sheet of cells, "strips", "rosettes", gland opening, "feathering"), and in the cells themselves. The cell size is uniform and enlarged. The cytoplasm is cyanophilic and occasionally vacuolated.

Examination of the sheets of cells does not reveal the typical honeycomb formation of normal endocervical epithelium due to crowding and overlapping of the nuclei. (Krumins et al., 1977; Bousfield et al., 1980; Gloor & Hurlimann, 1986; Betstil & Clark, 1987; Ayer et al., 1987; Pacey & Ng, 1997;Biscotti et al., 1997; Waddell,2003).

The columnar origin of cells can be recognized when lacunae, corresponding to glandular orifices, are present. At the edge of the sheets of cells, pseudo stratification of the nuclei may be observed. The glandular cells at the edge of a sheet are oriented with their long axis perpendicular to the edge. Some nuclei may have lost their surrounding cytoplasm and form irregular margins, resembling feathers at the edge of a bird's wing.

The smallest fragment is the case for 'strips' containing cells arranged in parallel with pseudo stratified nuclei and for 'rosettes', small round groups of cells with peripheral nuclei.

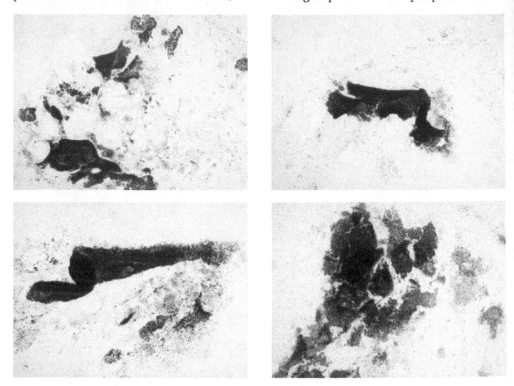

Fig. 6.3.1. AIS. Tightly crowded sheets, strips, rosettes, palisade, gland opening, feathering, of the malignant endocervical cells. The cell size is uniform and enlarged. Note crowding and overlapping of the nuclei. [Papanicolaou x 100 (all four fields)]

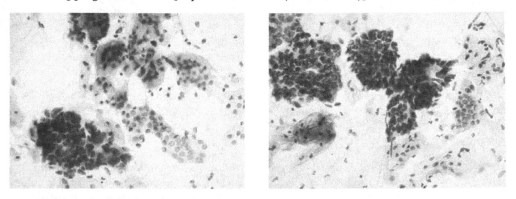

Fig. 6.3.2. AIS. Clusters of uniform small dark neoplastic cells and sheets of normal endocervical cells on the same field. (Papanicolaou x400)

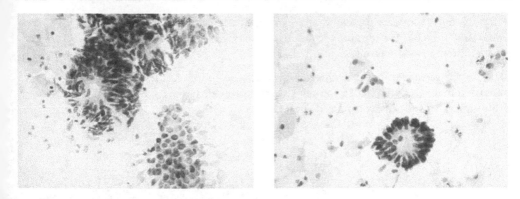

Fig. 6.3.3. AIS. Crowded sheet with "gland opening" and "rosette" of neoplastic cells. The nuclei are elongated, cigar-shaped, and hyperchromatic. Note a sheet of endocervical cells (left field) with slight nuclear enlargement and overlapping (GIL 1). (Papanicolaou x400)

The distinction between well differentiated and poorly differentiated AIS is based on nuclear features. In cell groups, the nuclei of cells of well-differentiated AIS are enlarged, oval or round, uniform, and have a regular nuclear membrane. When the cells are crowded the nuclei may be elongated, cigar-shaped, and hyper chromatic.

The chromatin is granular and evenly distributed. The nuclei in a portion of cells contain small nucleoli. Mitotic figures and apoptotic bodies are occasionally present.

Poorly differentiated AIS occurs less frequently then the well-differentiated type. (Fig.6.3.4.)

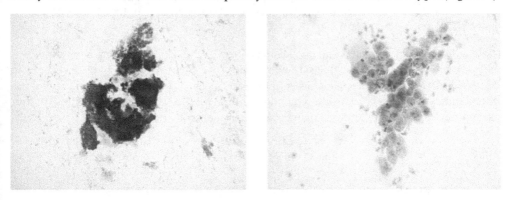

Fig. 6.3.4 Poorly differentiated AIS. The nuclei of the cells are round or irregular in shape and greatly enlarged, but less hyperchromatic and with finely granular chromatin. Nucleoli are multiple, irregular and enlarged. (Papanicolaou x100, x400).

In comparison to the cytological features of well-differentiated AIS, the nuclei of these cells are larger, but less hyperchromatic and with finely granular chromatin. The nuclei may be round and always contain nucleoli which may be multiple, irregular and/or enlarged. Mitotic figures can be seen. (Ayer et al., 1987; Pacey & Ng, 1997).

Although most endocervical adenocarcinoma in situ are of the usual 'endocervical' type, it is important to recognize that other variants sometimes occur. These include endometrioid, serous, and intestinal variants.

Of these, the most significant diagnostic variant is the endometrioid pattern (Lee, 1999). This pattern contains small cells in densely packed groups, having coarse nuclear chromatin exhibiting lack of pleomorphism (Fig.6.3.5.).

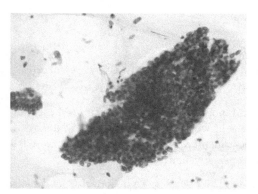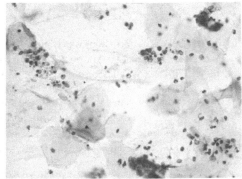

Fig. 6.3.5. AIS, endometrioid pattern, contains small cells in densely packet groups, having corse nuclear chromatin, exhibiting lack of pleomorphisam. Note apoptotic bodies on the right field (Papanicolaou x400).

These groups were more commonly misinterpreted as being of benign endometrial or endocervical origin (tubal metaplasia endometrioid variants). Criteria were developed to identify these cases as abnormal, at least to the level of atypical glandular cells: the absence of endometrial stromal cells and endometrial-like tubules, coarse chromatin patterns, extreme nuclear crowding, mitotic figures, and marginal feathering.

Key features of endocervical adenocarcinoma in situ are: hyperchromatic crowded groupings of cells, pseudostratified strips of columnar cells, epithelial rosettes, gland opening, nuclear and cytoplasmic 'feathering', twofold larger than normal nuclear size, endocervical nuclei, beyond normal increase of nucleus to cytoplasmic ratio, endocervical cells, coarsely granular and evenly distributed hyperchromatic chromatin, possible presence of small nucleoli, presence of mitotic figures and apoptotic bodies not associated with a background tumor diathesis

In a significant number of cases, abnormal squamous cells are present in association. (Fig.6.3.6) Focusing on these more commonly seen lesions can lead to a lack of identification of extant abnormal glandular processes. Careful observation and analysis of cervical cell samples should be exercised to identify these cells. Single malignant columnar cells may be mistaken as undifferentiated basal cells.

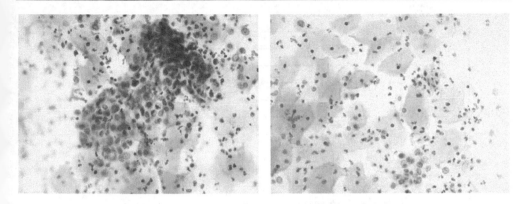

Fig. 6.3.6 CIS. A cluster of malignant squamous cells is present in association with small strip and single abnormal glandular cells (leading to lack of identification) (Papanicolaou x400).

7. Accuracy of cervical cytology

Intraepithelial lesions of the endocervical epithelium are more difficult to detect by cytology. However, cellular changes may frequently be less pronounced than those in squamous lesions and are difficult to observe unless architectural alterations call for attention. In mixed lesions, the glandular component may be eclipsed in abnormal cell count and intensity by the squamous component. (Boon et al., 1981; Di Tomasso et al., 1996).

In order to reach as accurate and precise a cytological diagnosis of intraepithelial lesions of endocervical cylindrical epithelium as possible, the cytological findings of patients with histologically verified adenocarcinoma in situ and mild to moderate glandular intraepithelial lesions were analyzed.(Ovanin-Rakic et al., 2010)

During the period 1993-2007, the value of cytology in the detection and differential diagnosis considering lesion severity and/or type of altered epithelium was assessed in 123 patients with a definite histological diagnosis of glandular lesions (AIS – n=13; GIL1 – n=11; GIL 2 – n7), glandular lesion associated with a squamous component (AIS+CIN/CI – n=58; GIL 1/GIL 2+CIN – n=28; GIL + MIC - n=6) (Table 2.).

Intraepithelial endocervical cylindrical lesions, with or without intraepithelial or invasive squamous component, were diagnosed in histological samples (78 biopsy specimens, 82 excochleation specimens, 70 conization specimens and 24 hysterectomy materials) from 123 patients aged 22-73 (mean 40).

The patients were divided into two categories : the first including 71 patients who were histologically diagnosed as AIS or AIS + CIN/CI, while the second included 52 patients who were histologically diagnosed as mild or moderate glandular intraepithelial lesions with squamous component (GIL1/ GIL2 + CIN, GIL + MIC) or without it (GIL1, GIL 2) .

In the first group, (table 2) cytological findings indicated epithelial abnormalities in 98.6% (70/71) patients. Considering lesion severity, the cytological and histological diagnoses were identical in 93% (66/71) patients.

Cytology	n	Histology					
		AIS		AIS + CIN	AIS + CI	AIS+CIN/CI	
		n	%	n	n	n	%
AIS	9	8	**61,5**	1		1	1,7
AIS + CIN	15	1	7,7	**12**	2	14	24,2
AC/Abnormal	6	3	23,1	3		3	5,2
AI + CIN	2			2		2	3,4
AI + CI	4			3	1	4	6,9
GIL + CIN	9			8	1	9	15,6
CIN	21			20	1	21	36,2
MIC	2			1	1	2	3,4
Abnormal	2			2		2	3,4
Inflammation	1	1	7,7				
Total	71	13 (100,0)		52	6	58	100,0
%	100,0	18,3				81,7	

Table 2. Cytohistologic correlation of either pure adenocarcinoma in situ (AIS) or a mixed AIS/squamous abnormality

The accuracy of cytological diagnosis according to lesion severity and type of epithelium was 92.3% (12/13) for glandular lesions and 56.9% (33/58) for mixed lesions.

In predicting the type of epithelium involved, the agreement between cytological and histological diagnosis was recorded in 61.5% (8/13) of histologically pure (AIS) and 20.7% (12/58) of mixed lesions (AIS + CIN / CI).

The accuracy of cytological identification of abnormalities of a particular type of epithelium, histologically diagnosed as either pure or mixed lesions, was 92.3% (12/13) and 96.6% (56/58) for cylindrical and squamous epithelium.

In the second group, (table 3), cytological findings indicated epithelial abnormality in 90.4% (47/52) patients. Considering lesion severity, the cytological and histological diagnoses were identical in 80.8% (42/52) patients.

The accuracy of cytologic diagnosis according to lesion severity and type of epithelium was 61.1% (11/18) for glandular lesions and 35.3% (12/34) for mixed lesions.

In predicting the type of epithelium involved, the agreement between the cytological and histological diagnosis was recorded in 22.2% (4/18) of histologically pure (GIL I) and 20.6% (7/34) of mixed lesions (GIL1,2 + CIN / MIC).

The accuracy of cytologic diagnosis according to lesion severity and type of epithelium was 61.1% (11/18) for glandular lesions and 35.3% (12/34) for mixed lesions.

In predicting the type of epithelium involved, the agreement between the cytological and histological diagnosis was recorded in 22.2% (4/18) of histologically pure (GIL I) and 20.6% (7/34) of mixed lesions (GIL1,2 + CIN / MIC).

Cytology	n	Histology								
		GIL I	GIL II	Total		GIL I + CIN	GIL II + CIN	GIL + MIC	Total	
		n	n	n	%	n	n	n	n	%
GIL I	4	4		4	22,2					
GIL I + CIN	8	2	2	4	22,2	3		1	4	11,8
GIL II+ CIN	5	1		1	5,6	1	3		4	1,8
AIS + CIN	4		2	2	11,1	1	1		2	5,9
GIL + MIC	2					1		1	2	5,9
CIN	24	1	1	2	11,1	8	10	4	22	64,6
Inflammation	5	3	2	5	27,8					
Total	52	11	7	18	100,0	14	14	6	34	100,0
%	100,0			34,6				65,4		

Table 3. Cytohistologic correlation of either pure gladular ysplasia (GIL) or a mixed GIL/squamous sbnormality

The rate of cytological identification of abnormalities of a particular type of epithelium, histologically diagnosed as either pure or mixed lesions, was 61.1% (11/18) and 100% (34/34) for cylindrical and squamous epithelium, respectively.

However, the fact that AIS patients are older than women with squamous CIS (Brown & Wells, 1986) and that the reverse is true for AI and CI could imply that the progression of GIL to AIS must be slower than the progression of CIN lesions to CIS.

In contrast, AIS should progress to AI significantly more rapidly than does CIS into CI That , indeed, seems to be the case. This would leave ample time for detection of glandular dysplasia, but not necessarily AIS (Plaxe & Saltzstein, 1999).

A coexisting SIL may obscure the presence of glandular lesion because abnormalities involving exclusively squamous components were quite frequently observed in the latter, either because of more distinct criteria and easier recognition, or due to more pronounced cellular lesions, or because of the predominant population of abnormal squamous cells, especially when extensive or high grade.

Historically, only sporadic cases of AIS were reported after it was first defined in 1953 by Friedell & McKay, 1953, who described only its histological appearance.

In the 1970s and 1980s, descriptive studies detailing the cytological criteria necessary for the prospective cytological diagnosis of AIS of the cervix uteri were published, increasing awareness and the diagnostic skill of cytologists.

For the cytologist intraepithelial glandular lesions pose possibly the greatest challenge in cervical sreening.

In a number of cases published over the period of 15 years, Papanicolaou smear screening detected a glandular abnormality before confirmation of AIS on cone biopsy or hysterectomy in 32- 79% cases. (Ayer et al., 1987; Azodi et al., 1999; Östör et al., 2000; Shin et al., 2002; Ovanin-Rakic et al., 2010).

Our observation has been that the number of AIS cases we identified has increased with time after our first identification in 1986.

The Papanicolaou smear in our patients had a sensitivity of 74.2% in detecting a glandular abnormality preoperatively. The cytological differential diagnosis of AIS showed a 61.5% and of GIL 1 22.2% accuracy. These results are similar to other reports (Ayer et al., 1987; Ioffe et al., 2003). Ioffe et al., 2003 have shown that the application of a semiquantitative system for the diagnosis of noninvasive endocervical glandular lesions results in better diagnostic reproducibility even in diagnostically problematic cases. Papanicolaou smear that includes adequate material from the transformation zone and endocervix can be a useful method for detecting precursor lesions of adenocarcinoma of the cervix. It bears remembering that cytology should not be recommended as the definitive diagnostic investigation for adenocarcinoma of the cervix uteri. If a clinician is suspicious of cancer during clinical examination, then he or she should proceed to colposcopy and biopsy regardless of the cytologic findings (Pacey & Ng, 1997).

8. Differential diagnosis

Cytological analysis of glandular lesion abnormalities in vaginal-cervical-endocervical (VCE) smears is associated with a number of diagnostic difficulties. Interpretation of the results must be based on scientific knowledge, meticulous training and experience, and demands dedication. To reach a definitive diagnosis utilizing cells that have desquamated freely from epithelial surfaces or cells that have been forcibly removed from various tissues, requires detailed examination of all available evidence. In this case, consideration must be given to both procedural aspects of the cytology laboratory and to changes that modify individual cells or cell groups.

There is overlap between the cytological criteria for various glandular lesions of the cervix, thus requiring more rigorous criteria for defining both benign and malignant cervical glandular lesions.

The emphasis is on criteria that discriminate among non-neoplastic conditions, benign neoplasms that may mimic malignancy, and malignant neoplasms that may pose as benign entities. Appropriate clinical data are certainly of great help in solving some of these diagnostic issues. Awarwness of cellular changes, together with pertinent clinical information , will prevent diagnostic errors.

8.1 Non-neoplastic lesions

8.1.1 Inflammatory changes in endocervical cells

Endocervical cells are usually present in small clusters, sheets, strips and pseudorosettes (small round groups of cells with peripheral cytoplasm), with minimal nuclear overlapping.

Hyperchromasia and mild anisonucleosis may be present in round or oval nuclei. Chromatin is finely stippled and may be smudged in the texture. Reactive/reparative atypical glandular cells are uniform in size and shape with small to medium-sized regular nucleoli and abundant cytoplasm. They are typically arranged in groups, rather than singly.

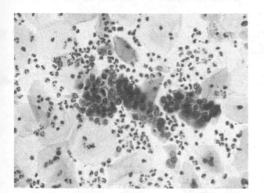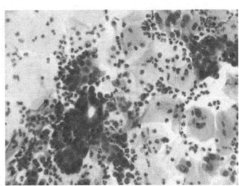

Fig. 8.1.1.1. A sheet of rective endocervical cells infiltrated by neutrophils with enlarged nuclei. Note a mitotic figure (left field) and gland opening (right field). These cells simulate glandular displasia.

The nucleoli are prominent, massive, spheroidal, and usually single, but they may vary in number. The cells can be multinucleated, variable in their size and shape, moulding, or overlapping each other, with very occasional mitoses seen in regenerating epithelial cells. There is a danger of mistaking these cells for endocervical adenocarcinoma cells. They differ by the regular distribution of their clumped chromatin and their smooth nuclear membrane. The most important feature which distinguishes sheets and clusters of endocervical cells with inflammatory changes from those of GIL is the exfoliation pattern. Nuclear stratification and feathering at the edge of sheets are features of GIL, which are rarely present in inflammatory smears. However, the cells seen in reactive conditions are usually monolayered with abundant cytoplasm. There is no stratification or 'feathering' of nuclei. Cells with marked nuclear enlargement, hyperchromasia and prominent nucleoli may be seen in polyps.

During pregnancy or the postpartum period, as a result of acute or chronic irritation, groups of endocervical cells can become considerably larger, with monstrous nuclei. They can be confused with anaplastic malignant cells, except for the persisting regularity of their smooth nuclear membrane and the abundance of their benign-appearing cytoplasm.

It is important to obtain clinical information in these situations and appraise cytological criteria for AIS with care (Naib, 1996; Pacey & Ng, 1997; Waddell, 2003).

8.1.2 Atypical repair

Reactive changes in epithelial cells are well described and generally well recognized as such by cytologists. Under some circumstances cells react to some injury of the epithelium. This condition of extreme reactivity, also known as atypical epithelial repair, can be problematic.

The cells seen in this condition may mimic a glandular abnormality, specifically invasive adenocarcinoma of the endocervix. (Fig. 8.1.2.1.) Cytoplasmic boundaries are well-defined and can clearly be seen in the overlapping or syncytial appearance of the groups noted in many neoplastic processes. Nuclei may be large with coarse chromatin and regular macro nucleoli are noted in virtually all nuclei. When repair becomes atypical, the nuclei begin to show variable degrees of pleomorphism of size and shape within the groups, often taking on nuclear contour irregularities. Chromatin patterns can turn from uniformly distributed to irregular and show coarse granularity. (Naib, 1996; Pacey & Ng, 1997; Waddell, 2003).

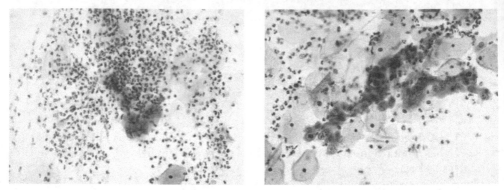

Fig. 8.1.2.1. This strips of atypical epithelial repair with enlarged nuclei. Some appear hyperchromatic and other have prominent nucleoli simulating adenocarcinoma. (Papanicolaou x400).

In determining diagnosis between atypical repair and invasive carcinoma, a designation of atypical glandular cells is warranted and an endocervical sampling procedure is required, as will be discussed below under management options.

8.1.3 Micro glandular endocervical hyperplasia

Micro glandular endocervical hyperplasia (MEH) is a localized proliferation of endocervical cells that can be mistaken for adenocarcinoma. MEH represents a non-neoplastic endocervical change usually related to progesterone effect or oral contraceptives. It is rarely seen in postmenopausal women.

The cytological manifestations of MEH falls in the spectrum of 'glandular atypia'. (Fig.8.1.3.1.)

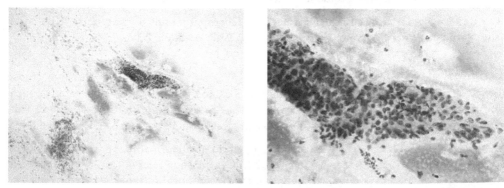

Fig. 8.1.3.1. A pseudostratified strip of endocervical cells is present. Nucleolar feathering at the periphery of the cluster, and nuclei are slightly enlarged. (Papanicolaou x 100, x 400).

The most common cytological findings are nuclear enlargement, nuclear hyperchromasia with fine nuclear chromatin, and nuclear overlap (Selvaggi and Haefner, 1997)

These are the presence of two- and three-dimensional fenestrated large sheets of cuboidal and columnar glandular cells, with finely vacuolated cytoplasm and with micro-rosette in sheets. Immature metaplastic cells, with dense basaloid cytoplasm, and reserve cells with little or no cytoplasm may also be seen. Reactive changes resulting in anisonucleosis, nuclear enlargement and prominent nuclei may lead to suspicion of either glandular or squamous neoplasia. The absence of chromatin heterogeneity, macro nucleolus formation, and tumour diathesis are the best discriminators in avoiding erroneous interpretation.

8.1.4 Sampling of the lower uterine segment, or post cone biopsy smears.

A cone biopsy shortens the endocervical canal allowing easier access to endometrial cells. Post cone biopsy smears may contain cells from this region which are referred to as lower uterine segment or LUS cells. Charateristic here is the presence of long tubular, branching glands embedded in loose monomorphic stroma. This is best observed on low power magnification.

Sampling of the lower uterine segment (LUS) following conisation is a result of endocervical brush or broom sampling of the endometrial cavity secondary to a shortened endocervical canal. This may occur following a conisation procedure or vigorous use of endocervical brushes in patients who have not undergone conisation. (Fig.8.1.4.1.)

Post cone biopsy smears are screened with a high index of suspicion, so the unwary can overreact to the presence of high endocervical cells or debrided endometrial cells in the smears.

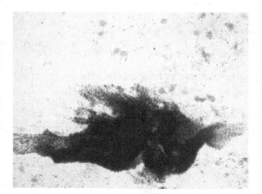 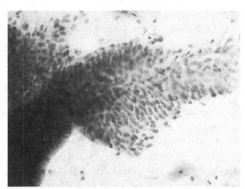

Fig. 8.1.4.1. Tightly crowded large group with pseudostratification and peripheral streaming of nuclei mimicking feathering (low uterine segment sampling) (Papanicolaou x100, x400).

Smears from the LUS show cellular two- or three-dimensional fragments with branching tubular glands that are embedded in stroma that is composed of round to spindle-shaped cells. In such circumstances, it is advisable to review the smear with the histology of the cone biopsy and with the previous abnormal Papanicolaou smear samples that led to the cone biopsy.

When compared with AIS smears, LUS sampling smears show smaller nuclei with less distinct nuclear membranes; densely dispersed, but finely granular chromatin, less frequent mitotic figures, and abundant endometrial-type stromal cells in the background (Hong et al., 2001). Presence of the endometrial-type stromal cells is significant; it is absent from the background of all cases of AIS.

Most false-positive interpretations are secondary to the presence of groups of nonciliated small glandular cells from either the upper endocervical canal or lower uterine segment of the endometrium (Lee, 1993, 1999).

These tightly crowded groups differ from AIS by having smaller, less hyperchromatic nuclei with finer chromatin, and by being intermixed with benign epithelial cells, and, occasionally, endometrial stromal cells. (Fig.8.1.4.2).

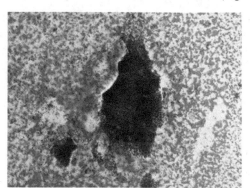 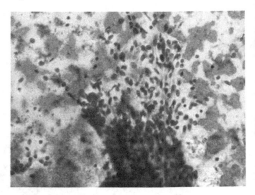

Fig. 8.1.4.2. Tightly crowded groups with benign epithelial cells, and, occasionally, endometrial stromal cells (post cone biopsy smears). (Papanicolaou x100, x400)..

More striking are the large and branching fragments of crowded glandular tissue.

Appreciation of the cytological features of LUS cells is essential to avoid misdiagnosis.

Presence of glandular cells of endometrial origin showing round nuclei, finely granular chromatin and nuclear crowding. Nucleoli are inconspicuous.

Occasional peripheral palisading of cells is noted and glandular openings are often visible. Presence of stromal cells showing uniform round to spindle-shaped nuclei, fine granular chromatin and scant cytoplasm. Peripheral cells are loosely attached and appear 'strung out'. It is important to exercise caution in examining post cone smears especially in women who have had a previous diagnosis of adenocarcinoma. Residual tumor may be present and careful scrutiny is required to differentiate abnormal from LUS cells.

8.1.5 Tubal metaplasia

Tubal metaplasia may pose a cytological problem. This refers to the replacement of normal endocervical glandular epithelium by foci of benign epithelium resembling that of normal fallopian tube epithelium. Apart from the smooth chromatin pattern of the nuclei, the most valuable feature for identification of tubal metaplasia is the density of cell cytoplasm, with blunted luminal edges bearing terminal bars and cilia. (Fig.8.1.5.1)

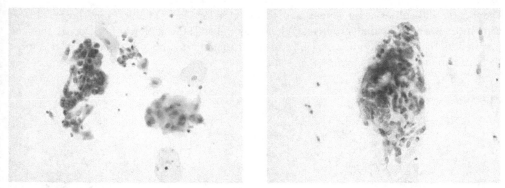

Fig. 8.1.5.1. Groups of cells with nuclear crowding, but the nuclear chromatin is finely granular. Note the terminal cilia. (Papanicolaou x400).

However, this may be a significant cause of false-positive smears for glandular neoplasia. When seen, it appears as flat sheets or cohesive clusters, and in palisade or mosaic patterns. It can mimic AIS because of nuclear crowding, nuclear overlap, nuclear feathering and nuclear palisading. Rosettes are uncommon, and, most importantly, the nuclear chromatin is finely granular and evenly distributed. Mitotic figures and apoptosis are rare. The identification of terminal bars and cilia is the most helpful cytological finding, but these features may not be present in some cases (Lee, 1993, 1999; Salvaggi & Haefner, 1997), since they may be lost during processing. Conversely, terminal bars and cilia are rarely seen in AIS.

The key to distinguishing difficult presentations of tubal metaplasia where cilia are absent is a careful review of nuclear chromatin. Most often the cells of tubal metaplasia will have normal 'endocervical' chromatin.(Fig.8.1.5.2.; Fig.8.1.5.3.; Fig 8.1.5.4.)

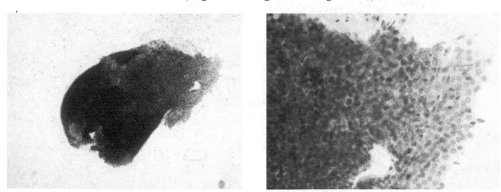

Fig. 8.1.5.2. Large three-dimensional and crowded groups of glandular cells. The cells at the edge of the fragments retain their cytoplasm and have a relatively smooth border and there is slight nuclear overlapping. (Papanicolaou x100, x400).

The typical chromatin pattern of AIS is coarse and evenly distributed. In addition, apoptotic nuclear fragments are not generally found in cases of benign tubal metaplasia, but may be commonly noted in neoplasias.

Diagnostic difficulties arise sometimes when cilia are not identified in large three-dimensional and crowded groups of glandular cells. Then, a borderline report may be justified as the possibility of coexistence of tubal metaplasia and glandular neoplasia must be borne in mind.

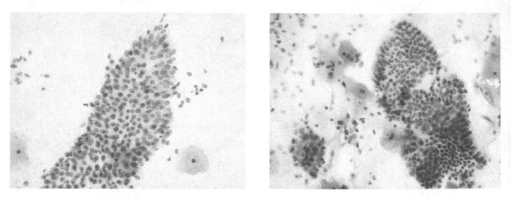

Fig. 8.1.5.3. Large groups of glandular cells with pseudo-stratification et the edge od the clusters. that mimicking feathering. The nuclear chromatin is finely granular and evenly distributed. (Papanicolaou x400).

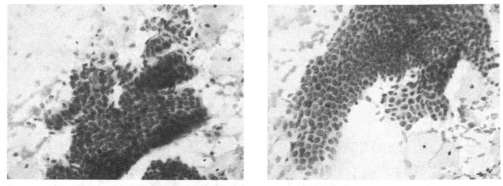

Fig. 8.1.5.4. . Large group and strips of glandular cells with palisading (left field). Large cluster that mimicking feathering (right field) . The nuclear chromatin is finely granular and evenly distributed. Note a mitotic figure on the right field. (Papanicolaou x400).

8.2 Neoplastic lesions

8.2.1 Carcinoma in situ

One of the major differential cytological diagnoses of AIS is endocervical gland involvement by CIS. In these cases, highly atypical nuclei are identified in the center of the cell aggregate, and some of the cells at the periphery of the aggregate appear to be endocervical cells. Involvement of endocervical glands by squamous CIS shows a syncytial arrangement, loss of cell polarity, and nuclear overlapping within the center of cell clusters whereas AIS cells generally maintain cell polarity. (Fig.8.2.1.1.)

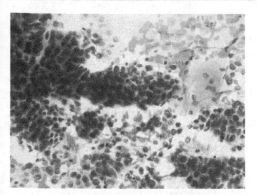 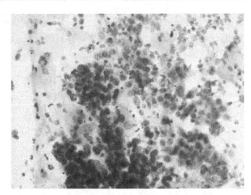

Fig. 8.2.1.1. CIS. A syncytial arrangement, loss of cell polarity, and nuclear overlapping within the center of cell clusters. Some of the cells at the periphery appear to be endocervical. (Papanicolaou x400).

The identification of occasional cells with dense eosinophilic cytoplasm and hyper chromatic cigar-shaped nuclei favors squamous CIS over AIS.

The identification of occasional cells with dense eosinophilic cytoplasm and hyper chromatic cigar-shaped nuclei favors squamous CIS over AIS.

Strips, rosettes, and gland formations, which are characteristic of AIS, are not observed in squamous CIS. In the infrequent cases that defy the above distinction, a diagnosis of atypical endocervical cells favouring AIS with a notation that squamous CIS cannot be excluded may be considered.

8.2.2 Adenocarcinoma in situ and invasive carcinoma

The most serious error is mistaking AIS for a benign process: small cell 'endometrioid' AIS, mistaken for direct sampling of the lower uterine segment endometrial cells; AIS mimicking tubal/tubo-endometrial metaplasia cells. This differential diagnosis may be extremely difficult or, in some cases, impossible in Papanicolaou smears. (Lee, 1993, 1999)

Endometrial adenocarcinoma may be mistaken for AIS if there is extension into the cervix and if the lesion is directly sampled.

Squamous carcinoma may be mistaken for adenocarcinoma if poorly differentiated.

9. New methods

A number of new technologies have been developed to improve the detection of cervical lesions, and a wide array of immuno-histochemical markers have been evaluated with respect to their specificity in staining abnormal cells in cervical cytological smears. However, there is still a significant demand for better biomarkers to identify neoplastic cervical glandular epithelial cells precisely. The most important advancement in cervical cytology has been the introduction of **liquid-based cytology (LBC)**. The advantages of LBC - compared to conventional cytology - are its increased sensitivity for detecting epithelial cell abnormality, reduced number of specimens with obscuring blood and inflammation, and the possibility of performing **molecular assay** directly from liquid-based specimens when a diagnosis of atypical cells is made (Bishop, 2002).

Human Papillomavirus (HPV DNA) detection is a potential biomarker of a neoplastic diagnosis in women with glandular abnormalities in their cervical smears. A positive HPV test is more strongly associated with squamous neoplasia than with glandular lesions.

Studies have shown that the prevalence of HPV in adenocarcinoma may be underestimated because the glandular epithelium does not support productive viral infections. HPV DNA in endocervical neoplasia is usually present in integrated form and not in the episomal particles. This integration may result in deletion of the viral genome. Detection of HPV DNA in the assay could depend on the presence of intact episomal HPV copies (Pirog et al., 2000).

Tumor suppressor protein (p16INK4a). Some studies have shown increased high-risk viral oncogene expression in dysplastic cervical epithelia, and have demonstrated that p16INK4a protein as a specific biomarker for the identification of dysplastic cervical epithelia in sections of cervical biopsy samples or cervical smears and in thin-layer LBC specimens (Murphy et al., 2002, Juric et al., 2006, 2010). The use of p16INK4a protein as a definitive marker for cervical neoplasia would be a valuable supplementary test in gynecologic cytology. A test result is considered positive if brownish granules are found in the nuclei and/or cytoplasm of dysplastic or malignant cells. (Fig.9.1.)

Murphy et al. 2004, compared the expression patterns of p16INK4a in benign and neoplastic glandular lesions and tubo-endometrioid metaplasia. All cases in each category displayed some p16INK4a expression.

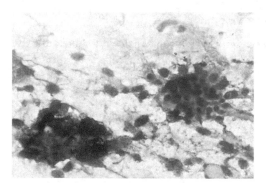 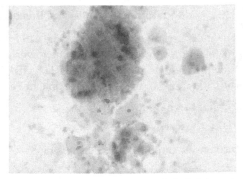

Fig. 9.1. p16INK4a postive staining of cluster malignant edocervical cells (AIS) left field, and of atypical endocervical cells (GIL2) and of high-grade squamous intraepthelial lesions (HSIL) on right field. Note a cluster of normal glandular cells p16INK4a negative staining on the left field. (x400, 100)

While p16INK4a has been demonstrated to be an excellent marker of cervical dysplasia in squamous neoplastic lesions of the cervix, it has potential pitfalls in cervical glandular lesions that may limit the utility of this biomarker in resolving the nature of suspicious glandular lesions, particularly in cytopathology.

Based on our results in detecting SIL lesions and carcinoma of the uterine cervix (Juric et al.,2010), immunocytochemical expressions of p16INK4a in ThinPrep cervical specimens correlate closely with the HPV-high-risk typed specimens through the polymerase chain reaction method (PCR) in the same samples.

We can assume that the combination of these tests can identify two groups within low-grade lesions, i.e. one with low risk for the development of premalignant cervical lesions, for which both of these tests are negative, and another group with both tests positive and with an increased risk of squamous intraepithelial lesions.

The value of immunocytochemical expressions of p16INK4a as adjunct methods for detection and differential diagnosis of glandular lesions has been investigated.

Imaging of silver-stained nucleolar organizer regions (AgNORs) is one of the more recent methods (Ploton et al. 1986). Nucleolar Organizer Regions (NORs) are structured from loops of ribosomal deoxyribonucleic acid (rDNA). Under the influence of RNA polymerase I, they are transcribed to ribosomes and proteins sited on the short arms of acrocentric chromosomes 13, 14, 15, 21, and 22. Since they have the central role in the transcription of nucleic acid into proteins, their number and size can be a reflection of cell proliferation, transformation or overt malignancy (Crocker, 1990). This method reveals AgNORs in the form of brown-black dots of different sizes within the nucleus. In numerous papers, the differential diagnostic and prognostic value of AgNOR analysis has been emphasized, on histological (Crocker, 1990; Darne et al., 1990) as well as cytological

(Fiorella et al., 1994; Audy i sur. 1995; Ovanin-Rakic & Audy-Jurkovic, 1998; Mahovlic et al., 1999) samples of benign, borderline and malignant lesions at various locations, and its significance has rarely been disputed.

Automated image analysis is applied to avoid the subjective error of an observer and to decrease the time necessary for data processing. This automated process was applied in 1996 as a fast, reproducible method on archival cytological specimens from cervix uteri stained by the Papanicolaou method (Ovanin-Rakic & Audy-Jurkovic, 1998) from 16 patients with a histological diagnosis (4 endocervical glandular dysplasia, 5 adenocarcinoma in situ, 7 adenocarcinoma invasivum) and 10 patients with benign endocervcal cells at the Institute of Gynecological Cytology, Department of Obstetrics and Gynecology, Medical School, University of Zagreb.

AgNORs are shown in the nucleus as dark brown to black dots. The count, area and size of AgNOR per square micrometer (minute <0.24; small 0.25 - 0.74; medium 0.75 - 1.4; large 1.5 - 2.4; extra large > 2.25) were analyzed in 50 cells per smears magnified 1,000x, on the focal plane. The SFORM system was used for digital image analysis (VAMS, Zagreb, Croatia) at the Institute of Pathology and Pathological Anatomy, Medical School, University of Zagreb. The system includes a high-resolution CCD color TV camera transferring images from the microscope (Olympus BHS, Tokyo, Japan) to a PC-compatible computer via a picture digitizer, with a resolution of 512 x 512 pixels, whereby each of them can assume a value described by 24 bits.

While measuring, the results of parameters measured are automatically transferred and logged in previously defined tables. The data obtained were processed on a PC by the

SPSS/PC+ 3.0 program (Chicago, Illinois, U.S.A.). Mann-Whitney and x2 tests were applied to test the differences between the groups, while statistical significance was tested at the level P= .05.

Our results showed that the mean values of AgNOR count and area per nucleus increased from benign endocervical cells (1.9; 2.17 µ2), and dysplasia (2.11; 2.53 µ2), and AIS (3.1; 3.27 µ2) to AI (3.,7; 5.49 µ2). The differences between all groups are statistically significant (P<.05)

Regarding AgNOR size and histological diagnosis, most frequently found were minute AgNORs in AIS (7.8%) and AI (6.7%), then benign (2,1%), and dysplasia (1.9%), while extra large AgNORs most frequently found in AI (15.9%). The differences between groups are statistically significant (P<.05) except for the pairs benign endocervical glandular cells and dysplasia.

One AgNOR per nucleus was usually present in benign endocervical cells (43.6%), and four or more in adenocarcinoma, especially adenocarcinoma invasivum (37.6%; 51.7%) with the differences between all groups being statistically significant (P<.05). (Fig.9.2.)

The AgNOR technique is a simple, inexpensive and reliable method applicable to both histological and cytological samples. AgNOR number is considered to be a reflection of cell proliferation. According to the literature, digital AgNOR image analysis of endocervical benign and abnormal glandular cells has not been performed before.

Our results indicate an increase in the mean value of AgNOR count from normal to intraepithelial and invasive glandular lesions, corresponding to the results on histological samples (Allen & Galimore, 1992; Darne et al., 1990; Miller et al., 1994), and cytological smears (Fiorella et al., 1994; Audy-Jurkovic et al., 1995). A significant finding of four or more AgNORs in 51,7% indicating adenocarcinoma invasivum that correlates to the results on histological samples (Miller et al., 1994).

Digital AgNOR image analysis (count, size and area) in cytological specimens of the cervix uteri indicated that the method is helpful in differentiating benign, intraepithelial and invasive lesions of the endocervical cylindrical epithelium, because statistically significant differences were obtained among all groups except for the benign state – dysplasia pair according to AgNOR size (p=0.8946).

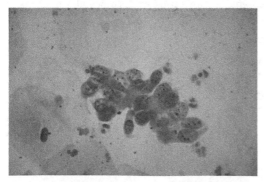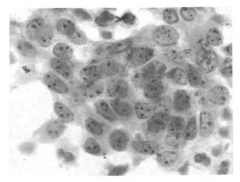

Fig. 9.2. AgNOR-stained. Cluster of adenocarcinoma in situ (left field), and adenocarcinoma invasivum (right fields). Note different types of brown-black dots within the nucleus.

10. Conclusion

Intraepithelial lesions of the endocervical epithelium are difficult to detect by cytology. However, recent studies show some favourable trends. In our study, the cytological differential diagnosis of AIS showed a 61.5% accuracy. The diagnostic accuracy of cytology is by far higher for pure (cylindrical only) than for mixed (cylindrical + squamous) lesions, because the abnormalities involving exclusively squamous component were quite frequently observed in the latter, either because of more distinct criteria and easier recognition, or due to more pronounced cellular lesions, or because of the predominant population of abnormal squamous cells.

The cytodiagnosis of cervical cylindrical epithelial lesions lags behind the cytodiagnosis of squamous epithelial lesions both in terms of screening and differential diagnosis. As data continue to accumulate, the clinical characteristics of pre-invasive glandular cervical lesions are becoming progressively better defined. Cytological screening for these lesions is imprecise. A major problem is the relative infrequency of glandular lesions and inexperience with sometimes difficult differentiation between benign glandular cells and the endocervix or lower segment of the endometrium. However, modifications to current classification systems may improve overall diagnostic accuracy. Nevertheless, all glandular abnormalities on the Papanicolaou smear require judicious evaluation and careful follow-up.

At present, the solution lies in better education. When in the hands of experienced cytologists, difficult cases of intraepithelial glandular lesions can be reliably distinguished from benign processes most of the time. The problem is in translating this experience to the entire community of cytologists, including cytotechnologists. Experience demands increased sensitivity, and cytologists and cytotechnologists both play a critical role in attempts to increase sensitivity in the face of demands for diagnostic specificity.

As our understanding of glandular lesions continues to expand and cervical sampling techniques continue to improve, we may expect continued enhancement in our ability to detect and treat intraepithelial glandular lesions, and thus help to decrease morbidity and mortality from cervical adenocarcinoma

11. References

Allen JP. & Gallimore AP. (1992). Nucleolar organizer regions in benign and malignant glandular lesions of the cervix. *The Journal of Pathology*, Vol.166, No.4 (February 1992), pp. 153-6, ISSN: 1096-9896

Audy-Jurković S. (1986). Citološka klasifikacija cerviksa uterusa. *Medicinska enciklopedija, II. Dopunski svezak*, Zagreb: Jugoslavenski leksikografski zavod, 93.

Audy-Jurković S, Singer Z, Pajtler M, Dražančić A & Grizelj V. (1992). Jedinstvena klasifikacija citoloških nalaza vrata maternice u Hrvatskoj. *Gynecol Perinatol*, Vol.1, No.4 (April 1992), pp. 185-8. ISSN: 1330-0091

Audy-Jurković S, Ovanin-Rakić A, Mahovlić V, Molnar Stantic B, Ilic-Forko J, Milicic D, Strelec M & Dimitrovski V. (1995). Citomorfologija argirofilnih nukleolarnih organizacijskih regija (AgNOR) u razlikovanju lezija endocervikalnih cilindričnih stanica. *Proceedings of the first Croatian congress of clinical cytology "Prvi hrvatski kongres kliničke citologije"*. Zagreb, March 1995

Ayer B, Pacey F, Greenberg M & Bousfield L. (1987). The cytologic diagnosis of adenocarcinoma in situ of the cervix uteri and related lesions. I. Adenocarcinoma in situ. *Acta Cytologica*, Vol.31, No.3, (May-June 1987), pp.397-411. ISSN: 0001-5547

Azodi M, Chambers SK, Rutherford TJ, Kohorn EI, Schwartz PE & Chambers JT. (1999). Adenocarcinoma in situ of the cervix: management and outcome. *Gynecologic Oncology*, Vol.73, No.3, (Jun 1999), pp. 348-53. ISSN: 0090-8258

Betsill WL,jr. & Clark AH. (1987). Early endocervical glandular neoplasia. Hystomorphology and cytomorphology. *Acta Cytologica*, Vol.30, No. 2, (March-April 1987), pp.115-26 ISSN: 0001-5547

Bertrand M, Lickrish GM & Colgan TJ. (1987). The anatomic distribution of cervical adenocarcinoma in situ implications for treatment. *American Journal of Obstetrics & Gynecolog*, Vol.157, No.1, (Juy 1987), pp. 21-25. ISSN: 0002-9378

Bharucha H, McCluggage G, Lee J & al. (1993). Grading cervical dysplasia with AgNORs using a semiautomated image analysis system. *Analytical & Quantitative Cytology & Hystology*,Vol15, No.5, (Octobar 1993), pp. 323-8. ISSN: 0884-6812

Biscotti CV, Gero MA, Toddy SM, Fischler DF & Easley KA. (1997). Endocervical adenocarcinoma in situ: an analysis of cellular features. *Diagnostic Cytopathology*, Vol.17, No.5, (November 1997), pp.326-32. ISSN: 8755-1039

Bishop JW. (2002). Cellularity of liquid-based, thin-layer cervical cytology slides. *Acta Cytologica*, Vol.46, No.4 (July-August 2002), pp.633-6. ISSN: 0001-5547

Boon ME, Baak JP, Kurver PJ & al. (1981). Adenocarcinoma in situ of the cervix: an underdiagnosed lesion. *Cancer*, Vol.48, No.3, (August 1981), pp. 768-773. ISSN: 0008-543x

Boon ME & Gray W(2003) Glandular Neoplasms of the Uterine Cervix, In: *Diagnostic Cytopathology*, Gray W & McKee GT, pp 651-7o5, Churchill Livingstone, ISBN 0 443 06473 3, China

Bousfield L, Pacey F, Young Q, Krumins I & Osborn R. (1980). Expanded cytologic criteria for the diagnosis of adenocarcinoma in situ of the cervix and related lesions. *Acta Cytologica*, Vol.24, No.2. (March-April 1980), pp. 283-96. ISSN: 0001-5547

Brown LJR & Wells M. (1986). Cervical glandular atypia associated with squamous intraepithelial neoplasia: a premalignant lesion? *Journal of Clinical Pathology*, Vol.39, No.1 (January 1986), pp. 22-8. ISSN: 0021-9746

Casper GR, Ostor AG & Quinn MA. (1997). A clinicopathologic study of glandular dysplasia of the cervix. *Gynecologic Oncology*, Vol.64, No.1, (January 1997), pp. 166-170. ISSN: 0090-8258

Chieng DC & Cangiarella JF. (2003). Atypical glandular cells. *Clinics in Laboratory Medicine* , Vol.23, No.5, (May 2003), pp. 633-7. ISSN: 0272-2712

Crocker J. (1990). Nucleolar organizer regions. *Current topics in pathology*, Vol.82, No.1, (January 1990), pp. 91 - 149. ISSN: 0070-2188

Darne JF, Polacarz SV, Sheridan E, Anderson D, Ginsberg R & Sharp F. (1990). Nucleolar organizer regions in adenocarcinoma in situ and invasive adenocarcinoma of the cervix. *Journal of Clinical Pathology*, Vol.43, No.8, (August 1990), pp. 657 - 660. ISSN. 0021-9746

DiTomasso JP, Ramazy I & Mody DR. (1996). Glandular lesions of the cervix. *Acta Cytologica*, Vol.40, No.6, (November-December 1996), pp. 1127-1135. ISSN: 0001-5547

Fiorela RS, Saran B & Kragel PJ. (1994). AgNOR counts as a discriminator of lesions of the endocervix. *Acta Cytologica*, Vol.38, No.3: (May-June 1994), pp. 527-30. ISSN: 0001-5547

Friedell GH & McKay DG. (1953). Adenocarcinoma in situ of the endocervix. *Cancer*, Vol.6, No.5, (September 1953), pp.887-97. ISSN: 0008-543x

Gloor E & Hurlimann J. (1986). Cervical intraepithelial glandular neoplasia (adenocarcinoma in situ and glandular dysplasia). A correlative study of 23 cases with histologic grading, histochemical analysis of mucins and immunohistochemical determination of the affinity for four lectins. *Cancer* (Phila), Vol.58, No.6, (September), pp. 1272-80. ISSN: 0008-543x

Goldstein NS, Ahmad E, Hussain M, Hankin RC & Perez-Reyes N. (1998). Endocervical Glandular atypia: Does a preneoplastic lesion of adenocarcinoma in situ exist? *American Journal of Clinical Pathology*, Vol.110, No.2, (August 1998), pp. 200-209. ISSN: 0002-9173

Hirschowitz Eckford SD, Phillpotts B & Midwinter A. (1994). Cytologic changes Associated with Tubo-Endometroid Metaplasia of the Uterine Cervix. *Cytopathology*, Vol.5, No.1, (February 1994), pp. 1 - 8. ISSN: 0956-5507

Higgins GD, Uzelin DM, Phillips GE, McEvoy P, Marin R & Burrell CJ. (1992). Transcription patterns of human papillomavirus type 16 in genital intraepithelial neoplasia: evidence for promoter usage within the E7 open reading frame during epithelial differentiation. Journal of General Virology, Vol.73, No.8, (August 1992), pp. 2047-57. ISSN: 0022-1317

Im DD, Duska LR & Rosenshein NB. (1995). Adequacy of conisation margins in adenocarcinoma in situ of cervix as a predictor of residual disease. *Gynecologic Oncology* Vol.59, No.2, (November 1995), pp. 179-82. ISSN: 0090-8258

Ioffe OB, Sagae S, Moritani S, Dahmoush L, Chen TT & Silverberg SG. (2003) Should Pathologists Diagnose Endocervical preneoplastic Lesions "Less Then" Adenocarcinoma In Situ?: Point. *International Journal of Gynecological Pathology*, Vol.22, No.1, (January 2003), pp. 18-21. ISSN: 0277-169

Juric D, Audy-Jurkovic S, Ovanin-Rakic A, Mahovlic V & Babic D. Introduction of p16 ink4a biomarker on fresh and archival cervical smears. (abstract). *Pathologica*, Vol.98, No.5, (October 2006) pp. 425.

Juric D, Mahovlic V, Rajhvajn S, Ovanin-Rakic A, Skopljanac-Macina L, Barisic A, Samija Prolic I, Babic D, Susa M, Corusic A & Oreskovic S. (2010). Liquid-based cytology - new possibilities in the diagnosis of cervical lesions. *Collegium Antropologicum*, Vol.34, N.1, (January 2010), pp 19 - 24. ISSN: 0350-6134

Krane JF, Lee KR, Sun D & Yuan L Crum CP. (2004). Atypical glandular cells of undetermined significance. Outcome predictions based on human papillomavirus testng. *American Journal of Clinical Pathology*, Vol.121, No.1 (January 2004), pp. 87-92. ISSN: 0002-9173

Krumins I, Young Q, Pacey F, Bousfield L & Mulhearn L. (1977). The cytologic diagnosis of adenocarcinoma in situ of the cervix uteri. *Acta Cytologica*, Vol.21, No.2 (March-April 1977), pp. 320-9. ISSN: 0001-5547

Kurian K & al-Nafussi A. (1999). Relation of cervical glandular intraepithelial neoplasia to microinvasive and invasive adenocarcinoma of the uterine cervix: a study of 121 cases. *Journal of Clinical Pathology*, Vol.52, No.2 (February 1999), pp.112-7. ISSN. 0021-9746

Kurman R & Solomon D. (1994) The Bethesda System for reporting cervical/vaginal cytologic diagnoses. Definitions, criteria, and explanatory notes for terminology and specimen adequacy. *Springer-Verlag*, New York ISBN: 0-387-94077-4

Lee KR. *(1993)*. Atypical glandular cells in cervical smears from women who have undergone cone biopsy. A potential diagnostic pitfall. *Acta Cytologica*, *Vol.37*, No.5 (September-October 1993), pp. 705-9. ISSN: 0001-5547

Lee KR, (1999). Adenocarcinoma in situ with a small (endometrioid) pattern in cervical smers: a test of the distinction from benign mimics using specific criteria. *Cancer Cytopathology*, Vol25. No.87 (October 1999), pp 254-8. ISSN 1934-662X

Ljubojevic N, Babic S, Audy-Jurkovic S, Ovanin-Rakic A, Jukic S, Babic D, Grubisic G, Radakovic B & Ljubojevic-Grgec D. (2001). Improved national Croatian diagnostic and therapeutic guidelines for premalignant lesions of the uterine cervix with some cost-benefit aspects. *Collegium Antropologicum*, Vol.25, N.2 (December 2001),*pp* 467-74. ISSN: 0350-6134

Mahovlic V, Audy-Jurkovic S, Ovanin-Rakic A, Bilusic M, Veldic M, Babic D, Bozikov J & Danilovic Z. (1999). Digital image analysis of silver-stained nucleolar organizer region associated proteins in endometrial cytologic samples. *Analytical & Quantitative Cytology & Histology* Vol.21. No.1 (January 1999),pp 47- 53. ISSN: 0884-6812

Miller B, Flax S, Dockter M & Photopulos G. (1994). Nucleolar organizer regions in adenocarcinoma of the uterine cervix. *Cancer*, Vol.15. No.73.(12) (december 1994), pp 3142 - *3145*. ISSN: 0008-543x.

Murphy N, Ring M, Killalea AG, Uhlmann V, O'Donovan M, Mulcahy F, Turner M, McGuinnes E, Griffin M, Martin C, Sheils O & O'Leary JJ. (2003). p16INK4A as a markea cervical dyskaryosis: CIN and GIN in cervical biopsies and ThinPrep smears of Clinical Pathology; Vol.56, No.1 (January 2003), pp.56-63. ISSN: 0021-9746.

Naib ZM. (1996). Cytology of the normal female genital tract. In: *Cytopathology, fourth edition* by Little, brown and Company. ISBN: 0-316-59674-4

National Cancer Institute Workshop. (1993). The revised Bethesda System for reporting cervical/vaginal cytologic diagnoses. *Acta Cytologica*, Vol.37, No.1, (January-February 1993), pp.115-24. ISSN: 0001-5547

National Cancer Institute Workshop. (1989). The 1988 Bethesda System for reporting cervical/vaginal cytologic diagnoses. Developed and approved at the National Cancer Institute Workshop. Bethesda, Maryland, USA, December 12-13. *Acta Cytologica*, Vol.33, No.4 (July-August 1989), pp. 567-74. ISSN: 0001-5547

National Cancer Institute Workshop. (1993). The revised Bethesda System for reporting cervical/vaginal cytologic diagnoses. *Acta Cytologica*, Vol.37, No.1, (January-February 1993), pp.115-24. ISSN: 0001-5547

NCI Bethesda System 2001. website http://bethesda 2001.cancer.gov

Nieminen P, Kallio M & Hakama M. (1995). The effect of mass screening on incidence and mortality of squamos and adenocarcinomaof the cervix uteri. *Obstet Gynecol*, VOL.85. No.6. (Jun 1995), pp. 1017-21. ISSN: 0029-7844

Östör AG, Duncan A, Quinn M & Rome R. (2000). Adenocarcinoma in situ of the uterine cervix: An experience with 100 cases. *Gynecologic Oncology*, Vol.79, pp. 207-210. ISSN: 0090-8258

Ovanin-Rakić A & Audy-Jurković S. (1988). Novije metode u citodijagnostici vrata maternice. In: Eljuga D, Dražančić A i sur. Prevencija i dijagnstika tumora ženskih spolnih organa. *Naknadni zavod Globus, Hrvatsko društvo ginekologa i opstetričara, Klinika za tumore i Hrvatska liga protiv raka, Zagreb*, 114-22. ISBN: 953-167-111-7

Ovanin-Rakić A, Pajtler M, Stanković T, Audy-Jurković S, Ljubojević N, Grubišić G & Kuvačić I. (2003). Klasifikacija citoloških nalaza vrata maternice "Zagreb 2002" Modifikacija klasifikacija "Zagreb 1990" i "NCI Bethesda System 2001 ". *Gynaecologia et Perinatologia*,Vol.12.No.4 (October-December 2003), pp 148-53. ISSN: 1330-0091

Ovanin-Rakic A, Mahovlic V, Audy-jurkovic S, Barisic A, Skopljanac-Macina L, Juric D, Rajhvajn S, Ilic-Forko J, Babic D, Folnovic D & Kani D. (2010). Cytology of cervical intraepithelial glandular lesions. *Collegium Antropologicum*, Vol.34, N.2 (Jun 2010), pp 401-406. ISSN: 0350-6134

Pacey NF, Ayer B & Greenberg M. (1988). The cytologic diagnosis of adenocarcinoma in situ of the cervix uteri and related lesions III. Pitfalls in diagnosis. *Acta Cytologica*, Vol.32, No.2 (March-April 1988), pp. 325-9. ISSN: 0001-5547

Pacey NF & Ng. ABP (1997). Glandular Neoplasms of the Uterine Cervix. Chapter 10. In: Bibbo M ed. *Comprehensive Cytopathology*. Philadelphia: Saunders.

Pajtler M & Audy-Jurković S. (2002). Pap smear adequacy: is the assessing criterion including endocervical cells really valid? *Collegium Antropologicum*, Vol.26, N.2 (December 2002), pp. 565-570. ISSN: 0350-6134

Park JS, Hwang ES, Park SN, Ahn HK, Um SJ, Kim CJ, Kim SJ & Namkoong SE. (1997). Physical status and expression of HPV genes in cervical cancers. Gynecologic Oncology, Vol.65. No.4 (April 1997), pp. 121-129. ISSN: 0090-8258

Pirog EC, Kleter B, Olgac S, Bobkiewicz P, Lindeman J, Quint WG, Richart RM & Isacson C. (2000). Prevalence of human papillomavirus DNA in different istological subtypes of cervical adenocarcinoma. American Journal of Pathology ,Vol.157. No.4 (October 2000), pp. 1055-1062. ISSN: 0002-9173

Plaxe SC & Saltzstein SL. (1999). Estimation of the duration of the preclinical phase of cervical adenocarcinoma suggests that there is ample opportunity for screening. *Gynecologic Oncology* Vol.75. No.1 (October 1999); pp. 55-61. ISSN: 0090-8258

Ploton D, Manager M, Jeannesson P, Himberg G, Pigeon F & Adnet JJ. (1986). Improvement in the staining and in the visualization of the AgNOR proteins (argyrophilic proteins of the

nucleolar organizer region) et the optical level. Histochemical Jurnal, Vol. 18 (, 1986),pp. 5-14. *ISSN: 1681-715x*

Rabelo-Santos SH, Derchain SFM, Amaral Westin MC, Angelo-Andrade LAL, Sarian LOZ, Oliveira ERZM, Morais SS & Zeferino LC. (2008). Endocervical glandular cell abnormalities in conventional cervical smears: evaluation of the performance of cytomorphological criteria and HPV testing in predicting neoplasia. *Cytopathology*, Vol.19. No.1 (February 2008) pp 34-43. ISSN: 0956-5507

Roberts JM, Thurloe JK, Bowditch RC & Laverty CR. (2000). Subdividing atypical glandular cells of undetermined significance according to the Australian modified Bethesda system. *Cancer,*Vol.90. No.2 (April 2000), pp. 87-95. ISSN: 0008-543x

Ruba S, Schooland M, Allpress S & Sterrett G. (1994). Adenocarcinoma in situ of the uterine cervix. Screening and diagnostic errors in Papanicolaou smears. *Cancer (Cancer Cytopathol)*, Vol.102. No.5 (October 1994) pp. 280-7. ISSN: 0008-543x

Selvaggi SM. (1994). Cytologic features of squamous cell carcinoma in situ involving endocervical glands in endocervical brush specimens. *Acta Cytologica*, Vol.38, No.4, (July-August 1994), pp. 687 - 692. ISSN: 0001-5547

Selvaggi SM & Haefner HK. *(1997)*. Microglandular hyperplasia and tubal metaplasia: pitfalls in the diagnosis of adenocarcinoma on cervical smears. *Diagnostic Cytopathology*, Vol.16. No.2 (February 1997) pp 168–73. ISSN: 8755-1039

Shin CH, Shorge JO, Lee KR & Sheets EE. (2002). Cytologic and biopsy findings leading to conization in adenocarcinoma in situ of the cervix. *Obstetrics & Gynecology*, Vol.100. No.2 (August 2002) pp 271-6. ISSN: 0029-7844

Singer A & Monaghan JM. (2000). Lower Genital Tract Precancer. Colposcopy, Pathology and Treatment. *Blackwel Science*, pp. 153-60 ISBN: 0-0632-04769-0

Solomon D, Davey D, Kurman R et al. (2002). The 2001 Bethesda System: terminology for reporting results of cervical cytology. *JAMA*, Vol. 287. No.16 (April 2002) pp 2114-9. ISSN: 0098-7484

Stoler MH, Rhodes CR, Whitbeck A, Wolinsky SM, Chow LT & Broker TR. (1992). Human papillomavirus type 16 and 18 gene expression in cervical neoplasias. Human Pathology, Vol. 23. No.2 (February 1992), pp. 117-28. ISSN: 0046-8177

Tase T, Okagaki T, Clark BA et al. (1989). Human papillomavirus DNA in glandular dysplasia and microglandular hyperplasia: Presumed precursors of adenocarcinoma of the uterine cervix. *Obstetrics & Gynecology*, Vol. 73. No.6 (June 1989) pp1005-1008. ISSN: 0029-7844

van Aspert - van Erp AJ, van t Hof-Grootenboer AB, Brugal G & Vooijs GP. (1995). Endocervical columnar cell intraepithelial neoplasia. *Acta Cytologica*, Vol.39, No.6, (November-December 1995), pp. 1199-1215. ISSN: 0001-5547

Waddell C. (2003) Glandular Neoplasms of the Uterine Cervix, In: *Diagnostic Cytopathology*, Gray W & McKee GT, pp 769-789, Churchill Livingstone, ISBN 0 443 06473 3, China

Willson C & Jones H. (2004). An audit of cervical smears reported to contain atypical glandular cells. *Cytopathology*, Vol.15. pp. 181-187. ISSN: 0956-5507

Zaino RJ. (2000). Glandular lesions of the uterine cervix. *Mod Pathology*, Vol.13. pp. 261-274. ISSN: 0893-3952

Part 2

Intraepithelial Neoplasia of Vulva

8

Current Insight into Specific Cellular Immunity of Women Presenting with HPV16-Related Vulvar Intra-Epithelial Neoplasia and Their Partners

Isabelle Bourgault-Villada

AP-HP, Hôpital Ambroise Paré, Boulogne Billancourt,
UVSQ, Versailles,
France

1. Introduction

The premalignant lesions of vulvar intraepithelial neoplasia (VIN) involve the mucosal and/or cutaneous epithelium of the vulva. VIN may be HPV-related VIN (usual VIN) or – unrelated and represents the most frequent vulvar cancer precursors. Usual VIN occurs in adult women and commonly resembles persistent anogenital warts which are often multifocal pigmented papular lesions. It is caused by high-risk HPV (HR-HPV) types, essentially 16 in up to 91% of the cases (Srodon et al, 2006), and histologically, it is made of poorly to undifferentiated basal cells and/or highly atypical squamous epithelial cells (McClugagge et al, 2009). The involvement of the entire thickness of the epithelium defines the grade 3 of the disease (VIN3). The disease progresses towards invasion in about 3% of treated patients and 9% of the untreated ones according to a review of over 3,000 cases (van Seters et al, 2005) whereas evolution towards invasive carcinoma is observed in about 30% of untreated grade 3 cervical intraepithelial neoplasia (CIN3) patients (Ostor et al, 1993).

2. Virology

HPVs are DNA viruses with a circular double strain genome including 8 000 base pairs. The genome is divided into three regions: a Long Control Region which controls viral replication, a region coding for Early proteins (E1 to E7, including the E6 and E7 proteins that share oncogenic and transforming properties), and a region coding for Late proteins such as L1 and L2 proteins that constitute 80% and 20% of the viral capside, respectively. More than 150 HPV have been sequenced, one HPV being considered different from another when there is a difference in 10% of nucleotides coding for L1 genes.

Following a breach in the malpighian pluristratified epithelium, HPVs infect basal stem cells of keratinocytes. The virus initially remains in episomal form with synthesis of E2 protein. This protein is a major regulator of viral vegetative cycle and is required for transcriptional regulation as well as viral DNA replication together with the E1 helicase (Desaintes et al, 1996). In contrast, E2 is generally undetectable in cancers due to a preferential integration of the viral genome in the cell genome and disruption of the E2 open reading frame (Berumen et al, 1994; Collins et al, 2009). Therefore E2 is a marker of viral infection and is specific for the early stages of the viral gene expression in infected cells. This was formally demonstrated in a recent work that showed a strong staining of the E2 protein in the intermediate differentiated layers of HPV16-infected tissues and low grade CIN (Xue et al, 2010). The high expression of HPV16 E2 in low grade lesions therefore represents a marker for HPV infection even before any clinical manifestation.

After integration of the genome of oncogenic HPVs such as HPV16 into the host genome, viral oncogenic E6 and E7 proteins are synthesized in large quantities in the inner third of the epithelium. E6 links to p53 and induces its degradation by the ubiquitin pathway and E7 links to pRB and allows the release of growth factors such as E2F.

During maturation of keratinocytes from the basal layer to the epithelial surface, viral capside proteins L1 and L2 are synthesized and expressed at the surface of mature keratinocytes in order to form a new viral particle which is able to infect adjacent healthy epithelium and to contaminate sexual partners.

3. Epidemiology of HPV16 related VIN

HPV infections occur preferentially in young women under 25 years of age (Boulanger et al, 2004). Several stages of lesions can be observed following oncogenic HPV infection. The first stage is a simple infection of keratinocytes that become koilocytes. The following stages are related to the transformation of infected keratinocytes into malignant cells. The depth at which malignant cells are found defines the disease stage. High grade squamous intraepithelial lesions as VIN3 are diagnosed on the basis of biopsy, with malignant cells in entire thickness of the epithelium

The premalignant lesions of HPV-related grade 3 intraepithelial neoplasia involve the mucosal and/or cutaneous epithelium of the vulva (usual VIN or VIN3), perineal and perianal region. Usual VIN occurs in adult women and commonly resembles persistent anogenital warts that are more often multifocal pigmented papular lesions disseminated on the vulva and/or the perianal skin than monofocal unique lesion (Figure 1).

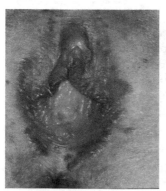 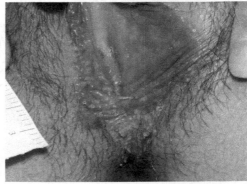

Current Insight into Specific Cellular Immunity of Women Presenting with HPV16-Related Vulvar
Intra-Epithelial Neoplasia and Their Partners

185

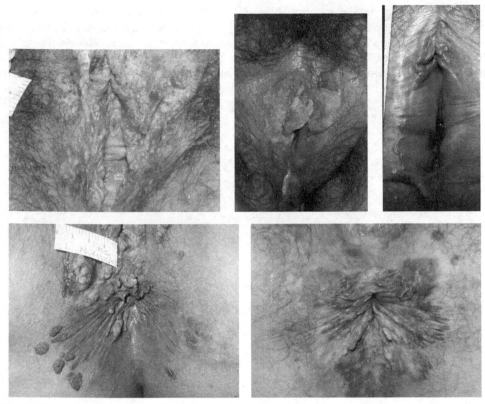

Fig. 1. Clinical presentations of usual multifocal or monofocal vulvar and preineal
intraepithelial neoplasia

4. Why does usual VIN can spontaneously regress?

Although usual VIN lesions are often chronic and recurrent, they can regress spontaneously
in up to 35% of young (less than 30 years) women presenting with multiple pigmented
lesions within a median duration of 9.5 months (Jones et al, 2005} (Bourgault Villada, 2010).
We previously studied a patient who presented with multifocal usual VIN and showed a
complete clearance of viral lesions eight months after disease onset and two months after
electrocoagulation of less than 50% of the usual VIN lesions (Bourgault Villada et al, 2004).
Immunohistochemical study of her initial vulvar biopsy revealed a marked dermal infiltrate
containing a majority of CD4+ T lymphocytes and an epidermal infiltrate made up of both
CD4+ and CD8+ T cells (Figure 2). She showed also a proliferating response against one
peptide from E6 protein and a high frequency anti-E6 and anti-E7 effector blood T cells by *ex
vivo* IFNγ– ELISpot assay just before clinical regression (Figure 3). Such a study of blood
cellular immune responses together with the analysis of vulvar biopsies obtained
simultaneously and correlated to clinical outcome was not previously reported. In an anti-
HPV vaccine trial conducted by Davidson and al (Davidson et al, 2003), usual VIN lesions
completely regressed in a patient following vaccination. Interestingly, immunostaining of

vulvar biopsy prior to the vaccine showed a marked CD4+ and CD8+ T lymphocyte infiltrate of both epithelial and sub-epithelial sheets. One may wonder whether the regression of these patient lesions could be related to a spontaneous regression. Therefore, the observation of a CD4+ and CD8+ infiltrate within sub-epithelial and epithelial sheets in the biopsy and the visualization of very strong blood anti-HPV T cell responses in patient with usual VIN could be predictive of spontaneous clinical outcome. It may also be thought that high numbers of blood CD4+ and CD8+ lymphocytes after therapeutic vaccination could allow clearance of HPV-16 lesions in usual VIN, assuming that anti-HPV vaccine-induced T effector cells could home in the HPV cutaneous and mucosal lesions.

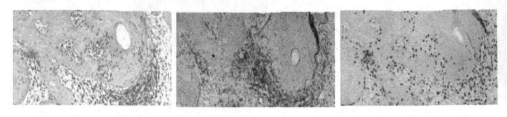

CD3 lymphocytes   CD4 lymphocytes   CD8 lymphocytes

Fig. 2. Immunohistochemical study of the vulvar biopsy just before spontaneous regression

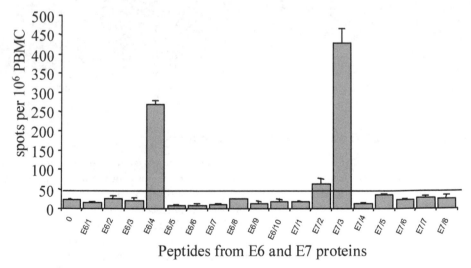

Fig. 3. IFNγ–ELISpot assay performed just before clinical regression

5. What is the exact role of cellular HPV16-specific T-cell responses?

Cellular immunity (CD4+ and CD8+ T-cells) plays a key role in the defense against all HPV-induced infections or lesions by destroying HPV-infected or -transformed keratinocytes. Indeed, the incidence of HPV infections and diseases significantly increases with CD4+ T cell impairment in immunosuppressed such as transplanted (Arends et al, 1997) or HIV-infected patients (Sun et al, 1997). In asymptomatic HPV16 infections, most women resolve spontaneously their infection without clinical disease concomitantly with blood anti-HPV16 Th1 CD4+ T cell responses (Welters et al, 2003). Similarly, regression of condyloma is associated with a dense epithelial cellular infiltrate made up of both CD4+ and CD8+ T lymphocytes with a Th1 cytokine profile as measured by cytokine mRNAs in interferon (IFN)-treated condylomas (Coleman et al, 1994). Proliferative CD4+ T cell responses are also associated with spontaneously regressive CIN3 (Kadish et al, 1997). The evolution of CIN3 towards invasive cancers is featured by a decrease of CD4+ cellular infiltrate, an increase of CD8+ T lymphocytes (Ghosh et al, 1992) with impairment of HPV16-specific CTL responses which could be related to a down- regulation of MHC class I molecules on HPV-16-infected cells and to the appearance of suppressive T lymphocytes (Treg) and a loss of blood anti-HPV-16 CD4+ activity. In high grade cervical intraepithelial neoplasia (CIN), positive intra-dermal reaction after intra-dermal injection of 5 HPV-16 E7 large peptides correlated with the spontaneous clearance of the lesions, which further indicates the presence and the very important role of HPV specific CD4+ T lymphocytes (Hopfl et al, 2000).

6. Anti-E2 T-cell responses are a marker of clinical viral control

We recently tested in a longitudinal study of 18 months, by proliferative assays, intracellular cytokines synthesis and IFNγ–ELISpot, the cellular immune responses against the HPV16 E2 protein that is early synthesized after HPV infection when the virus is episomal in eight women presenting with HPV16-related usual VIN and their healthy male partners (Jacobelli et al, 2011, unpublished data). In six women, we showed that anti-E2 polyfunctional CD4 T-cell responses (proliferative responses and synthesis of IFNγ and/or IL2) appear when the clinical lesions heal after treatment or when the HPV infection remains silent. In the women presenting with persistent lesions, no proliferation was observed.

Blood proliferative T-cell responses against HPV16 E2 peptides have been also observed in 50% of healthy women, who presumably previously cleared HPV16 infection (de Jong et al, 2004) and in 9 out of 22 regressive CIN3 cases (Dillon et al, 2007). In another studies, the lack of anti-E2 proliferative responses was reported in 16 of 18 patients (89%) affected with usual VIN lesions (Davidson et al, 2003) and in 7 of 8 and 9 of 12 women affected with CIN3 (Dillon et al, 2007; de Jong et al, 2004).These observations reinforce the strong role of T-cells in the control of HPV replication.

7. Why the male partners do not have any HPV16-related lesions?

Men are vectors of oncogenic HPV infection (Buckley et al, 1981; Giuliano et al, 2011). However, while HPV infection was found in 71 to 90% of the partners of HPV-infected women (Hippelainen et al, 1994; Nicolau et al, 2005), only 52% harbored the same HPV subtypes (Reiter et al, 2010). Moreover, penile intra-epithelial neoplasia is rare and detected in less than 2% of the men in contact with oncogenic HPV (Giraldo et al, 2008). We thus analyzed HPV infection and anti-HPV16 E2 blood T-cell responses in asymptomatic male partners chronically exposed to HPV16 during sexual intercourses with their wives affected with usual VIN (Jacobelli et al, 2011, unpublished data). We had hypothesized that male partners exposed to replicative HPV16 could develop immunologic responses against the early E2 viral protein and thus clear infection. In the absence of condom usage for at least 6 months, the male partners of women presenting with usual VIN could be contaminated by HPV16. HPV16 and HPV27 (a cutaneous HPV) were identified in genital sampling gathered by cytobrush in only two of the eight healthy partners. Such a prevalence of contamination by HPV16 is similar to the one usually observed in male partner of oncogenic HPV-infected women (Reiter et al, 2010). In this male population, we have chosen to study anti-HPV16 E2 T-cell responses because E2 protein is an early highly expressed protein. E2 is bigger than E6 and E7 proteins and induces more T-cell responses than E6 and E7. Therefore looking for an E2-specific response is then more sensitive. In addition, E2 is required for replication and the detection of E2-specific T cell responses is the signature of viral replication. The study of anti-E2 T-cell responses is then more appropriate for the early phases of HPV16 infection as supposed in male partners of women having usual VIN. We have observed HPV16-E2-specific proliferative responses in seven and intracellular cytokine synthesis of single IFNγ, dual IFNγ/IL2 and single IL2 in six out of the seven partners. Since there is no E2 protein in the viral particle, the high frequency of E2-specific T cells responses in partners of women with usual VIN demonstrates that the virus replicate in males.

These E2 specific T-cell responses indicate a striking correlation in all male partners but two between the absence of the HPV-related lesion. The presence of spontaneous E2-specific proliferative T-cell responses and single IFNγ, dual IFNγ/IL2, single IL2 T-cell producers was previously described in other viral systems (Harari et al, 2006; Pantaleo et al, 2006). These polyfunctional anti-E2 T-cell responses could be due to an efficient presentation of viral antigens by dendritic cells present in mucosal tissue and it is tempting to speculate that E2-specific responses are responsible for the clearing of the lesion. Therefore, spontaneously HPV control is related to the presence of memory polyfunctional CD4+ T-cells in male partners.

8. Why the prophylactic vaccine could be useful in men?

The analysis of E2 specific T cell responses is a sensitive and reliable tool to analyze disease progression and the natural history of HPV infection. In six out of eight male partners, the presence of T-cell proliferative responses and single IL2, dual IFNγ/IL2, single IFNγ memory T-cells against HPV16 E2 peptides was concomitant to the control of genital HPV lesions despite HPV16 exposure. These results are reminiscent of those described in Gambian prostitutes exposed to HIV with presence of anti-HIV cytotoxic T lymphocytes without any detectable HIV (Rowland-Jones et al, 1995). Such anti-viral immune T cells responses thus reflect an undetectable viral infection. Our experimental results demonstrate for the first time that, although not clinically detectable, HPV16 can replicate in men and can induce a strong memory T cell response against one of an early viral protein. The presence of polyfunctional (IL2, IFNγ/IL2 and IFNγ secretions and proliferation) anti E2 CD4+ T-cell responses in asymptomatic men unambiguously establishes that E2 is a marker of HPV infection even when undetectable lesions. Responses represent correlates of protective antiviral immunity in HPV infection. Monofunctional (production of IFNγ by IFNγ-ELISpot) "anti-E2 T-cell" responses does not allow HPV16 control. These results suggest that male are an important reservoir of HPV and provide a strong argument in favor of prophylactic HPV vaccination of young men with VLPs to decrease HPV16 infection in men, viral transmission from men to women and thus fight against the spread of mucosal HPV diseases in the population.

9. How to cure usual VIN? Therapeutic vaccines

Preventive vaccines do not address the current need for better treatment for women previously infected by HPV 16 or 18. Other types of vaccines must be used to increase or induce new specific anti-HPV cellular immunity (CD4+ and CD8+ T lymphocytes) in order to kill transformed epithelial cells. Several approaches can be used in this aim. To stimulate cytotoxic or antiviral CD8+ T lymphocytes, the vaccines must target the cytoplasm of dendritic cells. The degradation of vaccine antigens by proteasomes results in short peptides that can bind to HLA class I molecules and migrate at the surface of dendritic cells. To stimulate CD4+ T lymphocytes, endocytosis of vaccinal antigens is essential, followed by degradation of antigens by lysosome/endosome in large peptides that associate with HLA class II molecules before migrating at the surface of dendritic cells. All these therapeutic vaccines must target E6 and E7 viral proteins and contain recombinant viruses (vaccinia viruses for example), DNA or peptides.

Recently, an open clinical trial was performed by the Melief's group (Kenter et al, 2009) in twenty women presenting with usual VIN using 13 large peptides spanning the whole E6 and E7 proteins. Forty five percent of complete (9/20 women) and 25 % (5/20) of partial remission were observed 12 months after immunization. These important results would be even more interesting if the investigators had included a placebo group (Bourgault Villada, 2010a). A new trial with a placebo group is currently under way.

Vaccinia virus was also used in a recombinant vaccine containing E6 and E7 genes from HPV16 and HPV18 (TA-HPV) to vaccinate usual VIN patients. A clinical complete or partial response was observed in 8/18 treated women (Davidson et al, 2003). More recently, vaccination against usual VIN was also performed with another recombinant vaccinia virus, TA-L2E6E7 from HPV16 (Daayana et al, 2010). Two months before vaccination, 19 women were treated by topical imiquimod and then vaccinated by intramuscular route with 3 doses of recombinant vaccinia virus. Imiquimod is an immunomodulator that increases the synthesis of type I IFN by dendritic cells after its fixation to the TLR7 in human dendritic cells. Complete remission was obtained in 58% of vaccinated women.

10. How to determine the epitopic regions for a therapeutic vaccine?

In a study including 16 women presenting with usual VIN, we have determined the strongly immunogenic regions from HPV16 E6 and E7 proteins for CD4+ and/or CD8+ T lymphocytes (Bourgault Villada et al, 2010b). Among 18 large peptides of the proteins E6 and E7, two were recognized in proliferative assays as immunodominant by T cells from 10 out of 16 women (62%) at the entry in the study, namely E6/2 (aa 14-34) and E6/4 (aa 45-68) peptides. Four other peptides, E6/7 (aa 91-110), E7/2 (aa 7-27), E7/3 (aa 21-40) and E7/7 (aa 65-87) were recognized by only 12% of the women in proliferative or IFNγ–ELISpot tests. The regions of E6 and E7 proteins implicated in T cell recognition during HPV infection were not yet well defined because of the usually low frequency of anti-HPV blood T cell responses and of the difficulties of their study.

In protein E6, some peptides included in, including or overlapping our peptides E6/2 (aa 14-34) and E6/4 (aa 45-68) have already been described as preferentially recognized by CD4+ T cells. Among them, peptide E6 42-57 that is restricted by HLA-DR7 has already been identified (Strang et al, 1990). Regions E6 1-31, 22-51 and 24-45 can be also immunogenic for CD4+ T cells as shown in CIN or sexually active healthy women (Kadish et al, 1997). The region E6 42-71, which includes peptide E6/4 (aa 45-68), has also been described as a target of proliferative responses in CIN patients (Kadish et al, 1997). Another E6 111-158 region was previously described as inducing proliferative responses in infected asymptomatic subjects or in patients with CIN3 (Kadish et al, 1997; Strang et al, 1990) as well as E6 127-141 peptide in healthy young women (Gallagher et al, 2007). Similarly, peptides E7 43-77, E7 50-62 and E7 58-68 which are restricted by DR3, DR15 and DR17, respectively, were defined as epitopic peptides for CD4 + T cells (Strang et al, 1990; van der Burg et al, 2001; Wang et al, 2009). E7 region 51-98, including our E7/7 (aa 65-87) peptide, is also very immunogenic for proliferating T lymphocytes (de Gruijl et al, 1998; Luxton et al, 1996; Nakagawa et al, 1996).

The characterization of E6 and E7 HPV-16 epitopes and the HLA restriction of their recognition by CD8+ T lymphocytes are more precise: E6 29-38, E7 11-20, E7 82-90 and E7 86-93 epitopes are presented by HLA-A2 (Evans et al, 2001; Ressing et al, 1995, 1996), E6 80-88 and E7 44-52 by HLA-B18 (Bourgault Villada et al, 2000) and E6 49-57 by HLA-A24 (Morishima et al, 2007). In women who cleared HPV 16 infection, cytotoxic T lymphocytes (CTL) responses are directed against epitopes preferentially located in the N-terminal half of the E6 protein (region 16-40) (Nakagawa et al, 2005). In this fragment, the dominant epitope E6 29-37 is restricted by HLA-B48, E6 31-38 by HLA-B4002 and the subdominant epitope E6 52-61 by HLA-B35 (Nakagawa et al, 2007). The same group had also shown that the peptide E6 33-42 61 is recognized by CD8+ T lymphocytes in association with HLA-A68, peptide E6 52-61 in association with HLA-B57 and –B35, peptide E6 75-83 in association with HLA-B62, peptide E7 7-15 in association with HLA-B48 and peptide E7 79-87 in association with HLA-B60 (Nakagawa et al, 2004, 2007; Wang et al, 2008). In addition, E7 7-15 is also able to bind HLA-A2 and -B8 to be recognized by CTL (Oerke et al, 2005; Ressing et al, 1995). From the latter results, two hot spots of CD8+ T-cell epitopes in protein E6 may be located in the regions E6 29-38 and 52-61 and another one in protein E7 (E7 7-15) (Nakagawa et al, 2007). Nevertheless, a poor immunogenicity of E7 protein was observed in many studies during both HPV 16 infection and after peptidic vaccination using long peptides spanning both E6 and E7 (Kenter et al, 2008; Welters et al, 2008) such as those used in our study.

The epitopes E6/2 (aa 14-34) and E6/4(aa 45-68) hence could be strongly recognized by CD4+ and / or CD8+ T lymphocytes and could be particularly relevant in the design of a peptide vaccination. We may hypothesize that the T cell responses that we observed were able to contain the tumor cells into the epithelium. Therefore, E6/2 (aa 14-34) and E6/4 (aa 45-68) peptides could play a major role in the protection against invasive cancer by stimulating T lymphocytes. Specific CD4+ T-cells play an essential role in the defense against HPV in particular in women presenting with usual VIN and their male partners.

11. References

Arends, M. J., Benton, E. C., Mclaren, K. M., Stark, L. A., Hunter, J. A., & Bird, C. C. (1997). Renal allograft recipients with high susceptibility to cutaneous malignancy have an increased prevalence of human papillomavirus DNA in skin tumours and a greater risk of anogenital malignancy. *Br J Cancer* 75:722-8.

Berumen, J., Casas, L., Segura, E., Amezcua, J. L., & Garcia-Carranca, A. (1994). Genome
    amplification of human papillomavirus types 16 and 18 in cervical carcinomas is
    related to the retention of E1/E2 genes. *Int J Cancer* 56:640-5.
Boulanger, J. C., Sevestre, H., Bauville, E., Ghighi, C., Harlicot, J. P., & Gondry, J. (2004).
    [Epidemiology of HPV infection]. *Gynecol Obstet Fertil* 32:218-23.
Bourgault Villada, I., Beneton, N., Bony, C., Connan, F., Monsonego, J., Bianchi, A., Saiag, P.,
    Levy, J. P., Guillet, J. G., & Choppin, J. (2000). Identification in humans of HPV-16
    E6 and E7 protein epitopes recognized by cytolytic T lymphocytes in association
    with HLA-B18 and determination of the HLA-B18-specific binding motif. *Eur J
    Immunol* 30:2281-9.
Bourgault Villada, I., Moyal Barracco, M., Ziol, M., Chaboissier, A., Barget, N., Berville, S.,
    Paniel, B., Jullian, E., Clerici, T., Maillere, B., & Guillet, J. G. (2004). Spontaneous
    regression of grade 3 vulvar intraepithelial neoplasia associated with human
    papillomavirus-16-specific CD4(+) and CD8(+) T-cell responses. *Cancer Res* 64:8761-6.
Bourgault Villada, I. (2010a). Vaccination against HPV-16 for vulvar intraepithelial
    neoplasia. *N Engl J Med* 362:655-6.
Bourgault Villada, I., Moyal Barracco, M., Berville, S., Bafounta, M. L., Longvert, C., Premel,
    V., Villefroy, P., Jullian, E., Clerici, T., Paniel, B., Maillere, B., Choppin, J., & Guillet,
    J. G. (2010b). Human papillomavirus 16-specific T cell responses in classic HPV-
    related vulvar intra-epithelial neoplasia. Determination of strongly immunogenic
    regions from E6 and E7 proteins. *Clin Exp Immunol* 159:45-56.
Buckley, J. D., Harris, R. W., Doll, R., Vessey, M. P., & Williams, P. T. (1981). Case-control
    study of the husbands of women with dysplasia or carcinoma of the cervix uteri.
    Lancet 2:1010-5.
Coleman, N. H., Birley D, Renton, A. M., Hanna, N. F., Ryait, B. K., Byrne, M., Taylor-
    Robinson, D., & Stanley, M. A. (1994). Immunological events in regressing genital
    warts. *Am J Clin Pathol* 102: 768-74.
Collins, S. I., Constandinou-Williams, C., Wen, K., Young, L. S., Roberts, S., Murray, P. G., &
    Woodman, C. B. (2009). Disruption of the E2 gene is a common and early event in
    the natural history of cervical human papillomavirus infection: a longitudinal
    cohort study. *Cancer Res* 69:3828-32.
Daayana, S., Elkord, E., Winters, U., Pawlita, M., Roden, R., Stern, P. L., & Kitchener, H. C.
    (2010). Phase II trial of imiquimod and HPV therapeutic vaccination in patients
    with vulval intraepithelial neoplasia. *Br J Cancer* 102:1129-36.
Davidson, E. J., Boswell, C. M., Sehr, P., Pawlita, M., Tomlinson, A. E., Mcvey, R. J., Dobson,
    J., Roberts, J. S., Hickling, J., Kitchener, H. C., & Stern, P. L. (2003). Immunological
    and clinical responses in women with vulval intraepithelial neoplasia vaccinated
    with a vaccinia virus encoding human papillomavirus 16/18 oncoproteins. *Cancer
    Res* 63:6032-41.
De Gruijl, T. D., Bontkes, H. J., Walboomers, J. M., Stukart, M. J., Doekhie, F. S., Remmink,
    A. J., Helmerhorst, T. J., Verheijen, R. H., Duggan-Keen, M. F., Stern, P. L., Meijer,
    C. J., & Scheper, R. J. (1998). Differential T helper cell responses to human
    papillomavirus type 16 E7 related to viral clearance or persistence in patients with
    cervical neoplasia: a longitudinal study. *Cancer Res* 58:1700-6.
De Jong, A., Van Poelgeest, M. I., Van Der Hulst, J. M., Drijfhout, J. W., Fleuren, G. J., Melief,
    C. J., Kenter, G., Offringa, R., & Van Der Burg, S. H. (2004). Human papillomavirus

type 16-positive cervical cancer is associated with impaired CD4+ T-cell immunity against early antigens E2 and E6. *Cancer Res* 64:5449-55.

Desaintes, C., & Demeret, C. (1996). Control of papillomavirus DNA replication and transcription. *Semin Cancer Biol* 7:339-47.

Dillon, S., Sasagawa, T., Crawford, A., Prestidge, J., Inder, M. K., Jerram, J., Mercer, A. A., & Hibma, M. (2007). Resolution of cervical dysplasia is associated with T-cell proliferative responses to human papillomavirus type 16 E2. *J Gen Virol* 88:803-13.

Evans, M., Borysiewicz, L. K., Evans, A. S., Rowe, M., Jones, M., Gileadi, U., Cerundolo, V., & Man, S. (2001). Antigen processing defects in cervical carcinomas limit the presentation of a CTL epitope from human papillomavirus 16 E6. *J Immunol* 167:5420-8.

Fausch, S. C., Da Silva, D. M., & Kast, W. M. (2005). Heterologous papillomavirus virus-like particles and human papillomavirus virus-like particle immune complexes activate human Langerhans cells. *Vaccine* 23:1720-9.

Gallagher, K. M., & Man, S. (2007). Identification of HLA-DR1- and HLA-DR15-restricted human papillomavirus type 16 (HPV16) and HPV18 E6 epitopes recognized by CD4+ T cells from healthy young women. *J Gen Virol* 88:1470-8.

Ghosh, A. K., & Moore M. (1992). Tumour-infiltrating lymphocytes in cervical carcinoma. *Eur J Cancer* 28A: 1910-6.

Giraldo, P. C., Eleuterio, J., Jr., Cavalcante, D. I., Goncalves, A. K., Romao, J. A., & Eleuterio, R. M. (2008). The role of high-risk HPV-DNA testing in the male sexual partners of women with HPV-induced lesions. *Eur J Obstet Gynecol Reprod Biol* 137:88-91.

Giuliano, A. R., Lee J. H., Fulp, W., Villa, L. L., Lazcano, E., Papenfuss, M. R., Abrahamsen, M., Salmeron, J., Anic, G. M., Rollison, D. E., & Smith, D. (2011). Incidence and clearance of genital human papillomavirus infection in men (HIM): a cohort study. *Lancet* 377: 932-40.

Harari, A., Dutoit V., Cellerai, C., Bart, P. A., Du Pasquier, R. A., & Pantaleo, G et al. (2006). Functional signatures of protective antiviral T-cell immunity in human virus infections. *Immunol Rev* 211: 236-54.

Hippelainen, M. I., Yliskoski, M., Syrjanen, S., Saastamoinen, J., Hippelainen, M., Saarikoski, S., & Syrjanen, K. (1994). Low concordance of genital human papillomavirus (HPV) lesions and viral types in HPV-infected women and their male sexual partners. *Sex Transm Dis* 21:76-82.

Hopfl, R., Heim K., Christensen, N., Zumbach, K., Wieland, U., Volgger, B., Widschwendter, A., Haimbuchner, S., Muller-Holzner, E., Pawlita, M., Pfister, H., &Fritsch, P. (2000). Spontaneous regression of CIN and delayed-type hypersensitivity to HPV-16 oncoprotein E7. *Lancet* 356: 1985-6.

Jacobelli S., Sanaa. F.., Moyal Barracco M., Pelisse M., Berville S., Villefroy P., North M.O., Figueiredo S., Charmeteau B., Clerici T., Plantier F., Dupin N., Avril M.F., Guillet J.G., & Bourgault Villada I. (2011). Anti-HPV16 E2 protein T-cell responses and viral control in women with usual vulvar intraepithelial neoplasia and their healthy partners. *Submitted*.

Kadish, A. S., Ho, G. Y., Burk, R. D., Wang, Y., Romney, S. L., Ledwidge, R., & Angeletti, R. H. (1997). Lymphoproliferative responses to human papillomavirus (HPV) type 16 proteins E6 and E7: outcome of HPV infection and associated neoplasia. *J Natl Cancer Inst* 89:1285-93.

Kenter, G. G., Welters, M. J., Valentijn, A. R., Lowik, M. J., Berends-Van Der Meer, D. M., Vloon, A. P., Drijfhout, J. W., Wafelman, A. R., Oostendorp, J., Fleuren, G. J., Offringa, R., Van Der Burg, S. H., & Melief, C. J. (2008). Phase I immunotherapeutic trial with long peptides spanning the E6 and E7 sequences of high-risk human papillomavirus 16 in end-stage cervical cancer patients shows low toxicity and robust immunogenicity. *Clin Cancer Res* 14:169-77.

Kenter, G. G., Welters, M. J., Valentijn, A. R., Lowik, M. J., Berends-Van Der Meer, D. M., Vloon, A. P., Essahsah, F., Fathers, L. M., Offringa, R., Drijfhout, J. W., Wafelman, A. R., Oostendorp, J., Fleuren, G. J., Van Der Burg, S. H., & Melief, C. J. (2009). Vaccination against HPV-16 oncoproteins for vulvar intraepithelial neoplasia. *N Engl J Med* 361:1838-47.

Luxton, J. C., Rowe, A. J., Cridland, J. C., Coletart, T., Wilson, P., & Shepherd, P. S. (1996). Proliferative T cell responses to the human papillomavirus type 16 E7 protein in women with cervical dysplasia and cervical carcinoma and in healthy individuals. *J Gen Virol* 77 (Pt 7):1585-93.

Morishima, S., Akatsuka, Y., Nawa, A., Kondo, E., Kiyono, T., Torikai, H., Nakanishi, T., Ito, Y., Tsujimura, K., Iwata, K., Ito, K., Kodera, Y., Morishima, Y., Kuzushima, K., & Takahashi, T. (2007). Identification of an HLA-A24-restricted cytotoxic T lymphocyte epitope from human papillomavirus type-16 E6: the combined effects of bortezomib and interferon-gamma on the presentation of a cryptic epitope. *Int J Cancer* 120:594-604.

Nakagawa, M., Stites, D. P., Farhat, S., Judd, A., Moscicki, A. B., Canchola, A. J., Hilton, J. F., & Palefsky, J. M. (1996). T-cell proliferative response to human papillomavirus type 16 peptides: relationship to cervical intraepithelial neoplasia. *Clin Diagn Lab Immunol* 3:205-10.

Nakagawa, M., Kim, K. H., & Moscicki, A. B. (2004). Different methods of identifying new antigenic epitopes of human papillomavirus type 16 E6 and E7 proteins. *Clin Diagn Lab Immunol* 11:889-96.

Nakagawa, M., Kim, K. H., & Moscicki, A. B. (2005). Patterns of CD8 T-cell epitopes within the human papillomavirus type 16 (HPV 16) E6 protein among young women whose HPV 16 infection has become undetectable. *Clin Diagn Lab Immunol* 12:1003-5.

Nakagawa, M., Kim, K. H., Gillam, T. M., & Moscicki, A. B. (2007). HLA class I binding promiscuity of the CD8 T-cell epitopes of human papillomavirus type 16 E6 protein. *J Virol* 81:1412-23.

Nicolau, S. M., Camargo, C. G., Stavale, J. N., Castelo, A., Dores, G. B., Lorincz, A., & De Lima, G. R. (2005). Human papillomavirus DNA detection in male sexual partners of women with genital human papillomavirus infection. *Urology* 65:251-5.

Oerke, S., Hohn, H., Zehbe, I., Pilch, H., Schicketanz, K. H., Hitzler, W. E., Neukirch, C., Freitag, K., & Maeurer, M. J. (2005). Naturally processed and HLA-B8-presented HPV16 E7 epitope recognized by T cells from patients with cervical cancer. *Int J Cancer* 114:766-78.

Ostor, A. G. (1993). Natural history of cervical intraepithelial neoplasia: a critical review. *Int J Gynecol Pathol* 12:186-92.

Pantaleo, G., & Harari A. (2006). Functional signatures in antiviral T-cell immunity for monitoring virus-associated diseases. *Nat Rev Immunol* 6: 417-23.

Reiter, P. L., Pendergraft, W. F., 3rd, & Brewer, N. T. (2010). Meta-analysis of human papillomavirus infection concordance. *Cancer Epidemiol Biomarkers Prev* 19:2916-31.

Reiter, P. L., Pendergraft, W. F., 3rd, & Brewer, N. T. (2010). Meta-analysis of human papillomavirus infection concordance. *Cancer Epidemiol Biomarkers Prev* 19:2916-31.

Rowland-Jones, S., Sutton J., Ariyoshi, K., Dong, T., Gotch, F., McAdam, S., Whitby, D., Sabally, S., Gallimore, A., & Corrah, T. (1995). HIV-specific cytotoxic T-cells in HIV-exposed but uninfected Gambian women. *Nat Med* 1: 59-64.

Srodon, M., Stoler, M. H., Baber, G. B., & Kurman, R. J. (2006). The distribution of low and high-risk HPV types in vulvar and vaginal intraepithelial neoplasia (VIN and VaIN). *Am J Surg Pathol* 30:1513-8.

Strang, G., Hickling, J. K., Mcindoe, G. A., Howland, K., Wilkinson, D., Ikeda, H., & Rothbard, J. B. (1990). Human T cell responses to human papillomavirus type 16 L1 and E6 synthetic peptides: identification of T cell determinants, HLA-DR restriction and virus type specificity. *J Gen Virol* 71 (Pt 2):423-31.

Sun, X. W., Kuhn, L., Ellerbrock, T. V., Chiasson, M. A., Bush, T. J., & Wright, T. C., Jr. (1997). Human papillomavirus infection in women infected with the human immunodeficiency virus. *N Engl J Med* 337:1343-9.

Van Der Burg, S. H., Ressing, M. E., Kwappenberg, K. M., De Jong, A., Straathof, K., De Jong, J., Geluk, A., Van Meijgaarden, K. E., Franken, K. L., Ottenhoff, T. H., Fleuren, G. J., Kenter, G., Melief, C. J., & Offringa, R. (2001). Natural T-helper immunity against human papillomavirus type 16 (HPV16) E7-derived peptide epitopes in patients with HPV16-positive cervical lesions: identification of 3 human leukocyte antigen class II-restricted epitopes. *Int J Cancer* 91:612-8.

van Seters, M., van Beurden M., & de Craen, A. J (2005). Is the assumed natural history of vulvar intraepithelial neoplasia III based on enough evidence? A systematic review of 3322 published patients.*Gynecol Oncol* 97(2): 645-51.

Wang, X., Moscicki, A. B., Tsang, L., Brockman, A., & Nakagawa, M. (2008). Memory T cells specific for novel human papillomavirus type 16 (HPV16) E6 epitopes in women whose HPV16 infection has become undetectable. *Clin Vaccine Immunol* 15:937-45.

Wang, X., Santin, A. D., Bellone, S., Gupta, S., & Nakagawa, M. (2009). A novel CD4 T-cell epitope described from one of the cervical cancer patients vaccinated with HPV 16 or 18 E7-pulsed dendritic cells. *Cancer Immunol Immunother* 58:301-8.

Welters, M. J., de Jong A., van den Eeden, S. J., van der Hulst, J. M., Kwappenberg, K. M., Hassane, S., Franken, K. L., Drijfhout, J. W., Fleuren, G. J., Kenter, G., Melief, C. J., Offringa, R., & van der Burg, S. H. (2003). Frequent display of human papillomavirus type 16 E6-specific memory t-Helper cells in the healthy population as witness of previous viral encounter. *Cancer Res* 63(3): 636-41.

Welters, M. J., Kenter, G. G., Piersma, S. J., Vloon, A. P., Lowik, M. J., Berends-Van Der Meer, D. M., Drijfhout, J. W., Valentijn, A. R., Wafelman, A. R., Oostendorp, J., Fleuren, G. J., Offringa, R., Melief, C. J., & Van Der Burg, S. H. (2008). Induction of tumor-specific CD4+ and CD8+ T-cell immunity in cervical cancer patients by a human papillomavirus type 16 E6 and E7 long peptides vaccine. *Clin Cancer Res* 14:178-87.

Xue, Y., Bellanger, S., Zhang, W., Lim, D., Low, J., Lunny, D., & Thierry, F. (2010). HPV16 E2 is an immediate early marker of viral infection, preceding E7 expression in precursor structures of cervical carcinoma. *Cancer Res* 70:5316-25.

9

Expression of Vascular Endothelial Growth Factors VEGF- C and D, VEGFR-3, and Comparison of Lymphatic Vessels Density Labeled with D2-40 Antibodies as a Prognostic Factors in Vulvar Intraepithelial Neoplasia (VIN) and Invasive Vulvar Cancer

Robert Jach et al*,
Department of Obstetrics and Gynecology,
Jagiellonian University Medical College, Kraków,
Poland

1. Introduction

Vulvar cancer consists of 2.5-5% of all cancers of the female genital tract. Poland is a country with an average occurrence of this tumor. Most women suffer from this disease between the ages of 60 and 70 years. Recently conducted epidemiological studies indicate that the incidence of intraepithelial neoplasia and vulvar cancer is increasing particularly in women under 50 years of age. The epidemiological model of vulvar cancer in young women involves the role of sexually transmitted infections (sexually transmitted diseases), especially HPV infection with high oncogenic potential, such as HPV 16, 18, 45, 56, 66 and 69; also considered is the importance of habitual smoking in the process of disease genesis.

In older women, vulvar cancer coexists in a high percentage of cases with hyperplasia, lichen sclerosus and squamous cell carcinoma. In 60% of women, vulvar cancer develops in the labia majora, to a lesser percentage of the labia minora, the clitoris and posterior labial commissure. The method of choice in the treatment of vulvar cancer is surgery. Radiation and chemotherapy treatment are usually combined with surgical treatment. Radical excision of the vulva, together with regional lymph nodes, is an operation that involves early and late complications (Judson et al., 2006). Removal of lymph nodes in which metastatic cells are present leads to a reduction in tumor mass, which can have a positive therapeutic effect.

* Grzegorz Dyduch[2], Małgorzata Radoń-Pokracka[1], Paulina Przybylska[1], Marcin Mika[1],
Klaudia Stangel-Wojcikiewicz[5], Joanna Dulińska-Litewka[3], Krzysztof Zając[1], Hubert Huras[1],
Joanna Streb[4] and Olivia Dziadek[1]
[1] *Department of Obstetrics and Gynecology, Jagiellonian University Medical College, Kraków, Poland*
[2] *Department of Pathology, Jagiellonian University Medical College, Kraków, Poland*
[3] *Department of Medical Biochemistry Jagiellonian University Medical College, Kraków, Poland*
[4] *Department of Oncology, Jagiellonian University Medical College, Kraków, Poland*
[5] *Department of Gynecology and Oncology, Jagiellonian University Medical College, Krakow, Poland*

Cancer metastases present in the lymph nodes removed (lymphadenectomy) is also important for the final classification of the clinical stage of cancer and to decide on how to continue therapy.

Histopathologic assessment of lymph nodes removed in the vulvar cancer operation shows that metastatic cancer cells are found in the I ° stage of cancer in about 16%, grade II clinical stage in about 36%, an average of approximately 28 to 33% of cases. Thus, in approximately 70% of the women operated on with vulvar cancer, whereby there is no evidence of metastasis to the lymph nodes, removal does not improve treatment results (Markowska, 2006). Furthermore, the removal of normal lymph nodes may also have an adverse impact on the local immune status, which is important to the treatment of cancer.

Vulvar intraepithelial neoplasia (VIN) and its classification remains controversial. There are currently three systems of classification VIN:

1. The 3-staged, WHO classification system: VIN1-3
2. The Bethesda system type classification, two-staged, dividing the low-VIN and high degree VIN, and
3. The 2004 established ISSVD classification (International Society for the Study of Vulvovaginal Disease) does not stage VIN. The incidence of VIN has increased in recent decades, while the incidence of invasive vulvar cancer has remained at the same level or even declined in some countries. For example, the U.S. prevalence of VIN3 (vulvar carcinoma in situ) increased by 411% between 1973 and 2000, while the incidence of invasive vulvar cancer increased by only 20% during the same period of time (Judson et al., 2006).

Women with the immunodeficiency virus are approximately four times more vulnerable to HPV infection. However, the incidence of VIN in HIV-positive women ranges from 0.5 to 37% (Kuhn et al., 1999). Thus, a high percentage of HIV infection in women with VIN suggests recommended HIV testing for women with VIN. The lifetime risk of developing invasive cancer in women previously treated for VIN3 is between 2.5-7% (Iversen & Treli, 1998; Jones & Rowan, 1994; Thuis et al., 2000). (Figure 1).

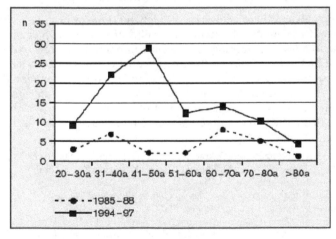

Fig. 1. VIN 2/3 incidence.

VIN is also an independent predictor of relapse (relative risk 3.06) as demonstrated in a study of 101 patients and 33 recurrences Preti (Preti et al., 2000).

In recent years, it has been observed that there is an increased incidence of vulvar cancer in women of younger ages leading to the search for less radical, but also effective surgical methods. These methods have allowed, on one hand, the reduction of injury, reduction of surgery time and reduction in the rate of postoperative complications and importantly, improvement in the quality of life for these women. To achieve this goal, it is equally important to recognize new prognostic factors, among which molecular factors are an attractive model, both in terms of diagnostics and therapeutic potential.

Lymphatic vessels play a key role in the spread of cancer. In recent years, several markers have been identified specific for lymphatic endothelium, which allowed for improved knowledge of the interaction between lymph vessels and lung cancer, but many issues in relation to their prognostic significance remains unclear (Judson et al., 2006; Markowska, 2006).

Proteins belonging to the family of glycoprotein endothelial growth factor (VEGF), referred to as VEGF-C and VEGF-D, are considered the most important regulatory factors in lymphangiogenesis. These factors are potent mitogens to the lymphatic and vascular endothelium. Furthermore, VEGF-C causes an increase in vascular permeability. These regulatory factors are ligands for the receptor VEGFR-3, whose expression is restricted to the endothelium of lymphatic vessels, and in the formation of blood vessels during embryogenesis. Factors VEGF-C and VEGF-D have been identified as stimulators of lymphatic endothelial proliferation, acting through the activation of the receptor 3 VEGF (VEGFR-3), which functions as a specific receptor in mature tissues and shows strong expression within the endothelial cells (Wissmann & Detmar, 2006; Najda & Detmar, 2006). Many clinical studies have shown a positive correlation between expression of VEGF-C and VEGF-D in the primary tumor and lymph node metastases (Donoghue et al., 2007; Nisato et al., 2003).

2. Aim

The aim of the study is to compare the immunohistochemical expression of vascular endothelial growth factors VEGF-C and D, and the expression of VEGFR-3, in VIN and vulvar invasive cancer, and to also compare the density of lymphatic marker D2-40 antibody in both groups, as prognostic factors, and to compare them with other clinicopathologic features.

3. Material and methods

The study was based on tissue material obtained during surgical procedures performed in the Department of Gynecology and Oncology at Jagiellonian University from 2006-2008. This tissue was in the form of cubes stored in paraffin, kept in the archives of the Department of Pathology. The clinical data of patients treated was obtained from the Department of Gynecology and Obstetrics. (Figure 2-6). The material was fixed in formalin on a routine basis. The analysis included 100 cases of vulvar dysplasia (30 - VIN I, 10 - VIN2, 60-VIN3) of which the average age of the patient was 65 years, and 10 cases of vulvar cancer of which the average age was 71 years. Patients were observed in the Gynecology Oncology Clinic of Jagiellonian University, Krakow from 12 to 48 months. At intervals of 4 - 6 months, patients were evaluated by history and physical examination, undergoing cytologic-colposcopic screening, and assessment of HPV DNA. In case of recurrence, a second operation was performed. The average age of women with vulvar cancer was 71.1 (59-79) years of age and was significantly higher than the average age of women with VIN, which was 56.6 (43-78) years at p = 0.003. (Table 1, Figure 7)

Type	n	Average mean	SD	min	max
VIN3	60	56,0	14,4	43	78
VIN2	10	67,0		67	67
ViN1	30	54,3	3,8	50	57
All groups VIN	100	56,6	11,5	43	78
Ca	100	71,1	7,2	59	79

Table 1. Charakteristics of women studied with VIN and vulvar cancer.

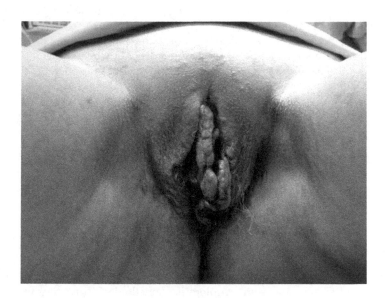

Fig. 2. VIN3 clinical manifestation.

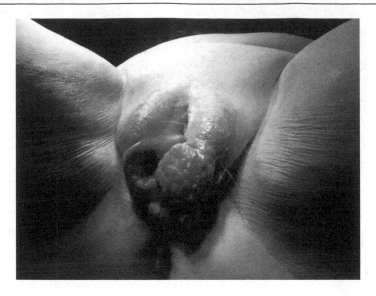

Fig. 3. Vulvar cancer. Clinical presentation.

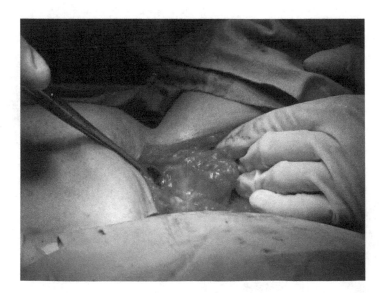

Fig. 4. Metastatic inguinal lymphnodes in vulvar cancer.

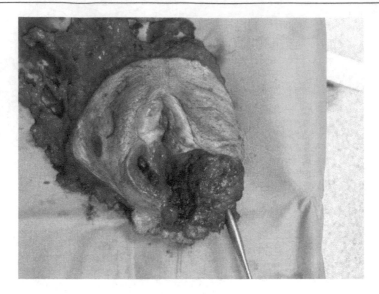

Fig. 5. Postoperative speciemen

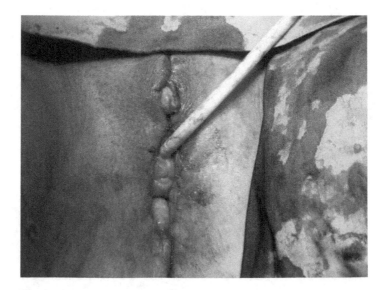

Fig. 6. Clinical presentation after surgery in vulvar cancer.

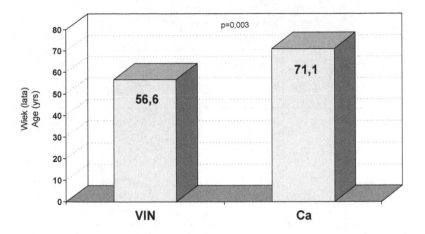

Fig. 7. Age distribution in VIN and vulvar cancer.

The diagnosis of VIN and vulvar cancer was based on histological evaluation of specimens taken from the vulva with guidance of the colposcope. In every case, specimens were also assessed in the presence of the HPV DNA test using Hybrid Capture II (DIGENE Corp.) with the material taken directly from the vulvar brush made of Dacron. The sample was also assessed histologically for metastasis in lymph nodes (superficial inguinal and deep) collected bilaterally in all 100 cases of vulvar cancer and 100 cases of VIN (clinically examined enlarged lymph nodes) during surgical treatment.

For immunohistochemical studies, samples were selected individually and made into paraffin blocks, each representative of a specific case.

From the selected paraffin cubes, 4 microns thick, samples were placed on glass slides coated with silanized basic Super Frost + (SuperFrost Inc.). Deparafinization involved placing the sample for 10 minutes in xylene and then dehydrated lead through three changes of ethyl alcohol of increasing concentration (70%, 86% and 96%), each lasting 5 minutes. In order to inhibit endogenous peroxidase activity, preparations were placed for 10 min in 3% H_2O_2 solution. Antigen unmasking was achieved by heating in a microwave oven Whirlpool, 3 times for 5 minutes in a 750W preparation placed in citrate buffer (pH 6.0, 0.01m), or EDTA buffer (pH 8.0, 0.01M).

After incubation, preparations were washed with TBS buffer (50 mM Tris-Hcl, 150 mM NaCl, pH 7.6, DAKO Corporation). To visualize the antigen-antibody complex, we used the En Vision system (DAKO Corporation) and Lab Vision (LabVision) (see the table) with 3-amino-9-ethylcarbazole (AEC) (DAKO Corporation) as a chromogen. Nuclei were contrasted with *Mayer* Hematoxylin for 1 minute and then covered with slide coverslips in glycerol. Basic data on the antibody used in the work is presented in Table 2.

Positive control preparations were: tonsil - for D2-40, placenta- for VEGFR-3 and VEGF-C and VEGF-D, ductal carcinoma of the breast - for VEGF-C and VEGF-D, the small intestine - for D2-40. Negative controls were the same antibody preparations as the original.

Antibody	Type	Clone	Manufacturer	Dilution	Unmasking	Time	Detection system
VEGFR-3	monoclonal	KLT9	Novocastra	1:50	Microwave EDTA, pH=8,0	60 min	Lab Vision
VEGF-C	polyclonal		Santa Cruz	1:100	Microwave EDTA, pH=8,0	12 hrs	En Vision
VEGF-D	monoclonal	78923	R&D systems	1:200	Microwave EDTA, pH=8,0	12 hrs	En Vision
D2-40	monoclonal	D2-40	Covance	Ready-to-use	Microvave citrate buffer ph=6,0	30 min	En Vision

Table 2. Antibodies, their dilution and incubadion time

4. Evaluation of immunohistochemistry

Lymphatic vessel density was evaluated using high power (40x). (Figure 8-10) All D2-40 positive vessels within the area of 0,5mm width beneath the dysplastic epithelium (VIN) or invasive edge of the tumor were counted and the result has been provided. All vessels were counted and the result is given as the number of vessels per 2mm. (Figure 11, 12).

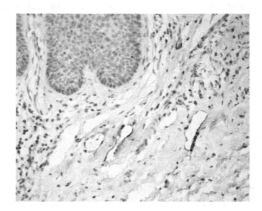

Fig. 8. Histologic picture of VIN I. Immunohistochemical staining with D2-40 antibody. Magnification 40x.

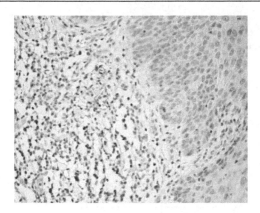

Fig. 9. VIN II . Small lymphatic vessels. Immunohistochemical staining with D2-40
antibody. Magnification 40x.

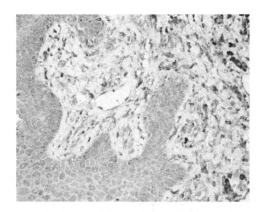

Fig. 10. Histologic picture of VIN. Strong VEGF-D expression in dysplastic epithelium.
VEGF-D positive, magnification 40x.

The staining for VEGF-C, VEGF-D, and VEGFR-3 expression was assessed by
semiquantitative method, at high (400x) magnification. The severity of expression was
evaluated on a scale from 0 to 3; 0-lack of staining, 1-weak, 2-moderate, 3-strong). The
number of cells expressing VEGF-C and D and VEGFR-3 were classified to four groups: 0 -
no staining or staining in individual cells at the edge of the preparation, 1 - less than 20%
positive cells, 2 - 20-50%, 3 - above 50%. Points earned for staining intensity and number of
positive cells finally summed 4 groups:

    0: 0-1 points
    I: 2-3 points
    II - 4 points
    III - 5-6 points

5. Results

The results are presented in Table 3-10.

DGN	n	VEGF-C	
		0	1
VIN3	60	30	30
		50,00%	50,00%
VIN2	10	10	0
		100,00%	0,00%
VIN1	30	30	0
		100,00%	0,00%
All Grps VIN	100	70	30
		70,00%	30,00%
CA	100	90	10
		90,00%	10,00%

Table 3. Expression of VEGF-C in studied groups of women.

DGN	n	VEGF-D			
		0	1	2	3
VIN3	60	10	20	30	0
		16,67%	33,33%	50,00%	0,00%
VIN2	10	0	10	0	0
		0,00%	100,00%	0,00%	0,00%
VIN1	30	20	1	0	0
		66,67%	33,33%	0,00%	0,00%
All Grps VIN	100	30	40	30	0
		30,00%	40,00%	30,00%	0,00%
CA	100	0	30	60	10
		0,00%	30,00%	60,00%	10,00%

Table 4. Expression of VEGF-D in groups of women studied.

DGN	n	VEGFR-3			
		0	1	2	3
VIN3	60	0	0	60	0
		0,00%	0,00%	100,00%	0,00%
VIN2	10	0	0	0	10
		0,00%	0,00%	0,00%	100,00%
VIN1	30	0	20	10	0
		0,00%	66,67%	33,33%	0,00%
All Grps VIN	100	0	20	70	10
		0,00%	20,00%	70,00%	10,00%
CA	100	10	0	30	60
		10,00%	0,00%	30,00%	60,00%

Table 5. Expression of VEGFR-3 in the studied groups of women.

Identification	n	D2-40			
		Average mean	SD	min	max
VIN3	6	4,88	0,81	4,0	6,0
VIN2	1	3,90		3,9	3,9
VIN1	3	2,00	0,00	2,0	2,0
All Grps VIN	10	3,92	1,49	2,0	6,0
Ca	10	3,80	1,76	1,7	7,0

NS

Table 6. The density of D2-40 vessel in the examined groups of women.

Identification	n	Recurrence		
		NO	YES	
VIN3	6	2	4	
		33,33%	66,67%	
VIN2	1	1	0	
		100,00%	0,00%	
VIN1	3	3	0	
		100,00%	0,00%	
All Grps VIN	10	6	4	
		60,00%	40,00%	
CA	10	7	3	
		70,00%	30,00%	p=0,639

Table 7. Presence of recurrences in examined groups of women.

| Indentification | n | Metastasises to lymph nodes | |
		NO	YES
VIN3	6	6	0
		100,00%	0,00%
VIN2	1	1	0
		100,00%	0,00%
VIN1	3	3	0
		100,00%	0,00%
All Grps VIN	10	10	0
		100,00%	0,00%
CA	10	8	2
		80,00%	20,00%

p=0,136

Table 8. Presence of metastasises in examined groups of women.

| Identification | n | HPVDNA | |
		NO	Yes
VIN3	6	2	4
		33,33%	66,67%
VIN2	1	1	0
		100,00%	0,00%
VIN1	3	2	1
		66,67%	33,33%
All Grps VIN	10	5	5
		50,00%	50,00%
CA	10	6	4

p=0,653

Table 9. Presence of HPV DNA in examined groups of women.

The statistical analysis (statistical package Statistica 8.0, Statsoft. Inc. USA) showed no significant differences in the expression of VEGF-C and D and VEGFR-3 between the VIN group and invasive vulvar cancer group. Weak expression of VEGF-C was found only in two cases of the analyzed series, and in all cases, the expression of VEGF-D and VEGFR-3 was observed. The strongest expression of VEGF-D and VEGFR-3 was observed in the group of invasive cancers. Similarly, the differences in the amount of lymphatic vessels between the group of invasive cancers and VIN group did not reach statistical significance.

The highest density of lymphatic vessels per 2 mm was observed in VIN. In this group, and in most cases, sections of lymphatic vessels were irregular and slightly expanded. In the cancer group, we observed small lymphatic vessels with a narrow, oval lumen. Moreover, in two cases, the presence of lymphovascular space invasion (LVSI) was observed.

The evaluation of recurrence in women treated showed a statistical difference between the group VIN3 and VIN1 at p = 0.058 (ie. 5.8%, with the conclusion that VIN3 and VIN1 differ by the presence of recurrence), which is an interesting trend and needs further investigation in more cases, with specific attention to the literature citing similar occurrence of relapses and VIN1 and VIN2 / 3 (Table 7).

The median survival time without recurrence was the longest in the group of patients with vulvar cancer, in which it was followed by 45.2 months; in women with VIN, it was followed by 37.2 months The shortest survival time without recurrence was among women with VIN3-28.7 months, 95% CI (Table 10). In our opinion, this trend may result from the heterogeneity of women diagnos ed with VIN3, location and multiplicity of the disease.

Group	Median survival time without recurrence	95% CI (95% Confidence Interval)
VIN3	28,7 mos.	14,4 mos. – 43,0 mos.
VIN	37,2 mos.	25,8 mos. – 48,6 mos.
Ca	45,2 mos.	37,6 mos. – 52,8 mos

Table 10. Comparison of average survival time without recurrence in the treated groups of patients.

Disease-free survival curves were compared using log-rank tests. There were no statistically significant differences between survival without recurrence in groups of Ca. and VIN. Two-year survival among patients with VIN was 60%, and for patients with vulvar cancer, 70% (Figure 13). No statistically significant differences in the prevalence of HPV DNA in the test groups and the presence of lymph node metastases in the groin were observed. (Table 8, 9).

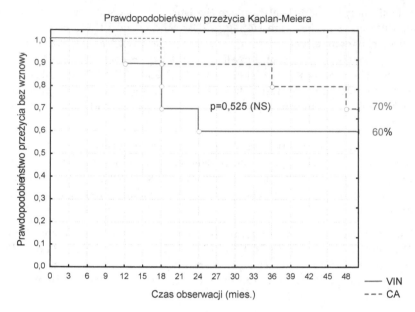

Fig. 13. Recurrence- free survival rate in cancerous and VIN patients.

6. Discussion

Lymphatic vessels play a key role in the spread of cancer. In recent years, several markers specific to lymphatic endothelium have been identified, which allowed for a deeper insight into the relationship between lymph vessels and cancer, although there are still a lot of questions for which there is no clear answer (Hillen & Griffioen, 2007; Sundar & Ganesan, 2007).

The idea that cancer cells spread to already existing lymphatic vessels is still present in the literature, although it seems that there is a belief about the presence of active lymphangiogenesis in tumors (Beasley et al., 2002; Maula et al., 2003; Nathanson, 2003). Currently, though one does not deny the presence of lymphatic vessels in tumors, the problem of the existence of active lymphangiogenesis, and the prognostic value of lymphatic vessel density inside or at the periphery of the tumor has not been completely resolved (Nathanson, 2003; Stacker et al., 2001; Ji, 2006). The functionality of the lymphatic vessels, and their role in metastasis is the subject of discussion (Maula et al., 2003; Ji, 2006; Padera et al., 2002).

In the presence of primary tumors, the extent of lymphangiogenesis can serve as a prognostic indicator of survival. A positive correlation is observed in the density of lymphatic vessels with lymph node metastasis; this has been reported in cases of squamous cell carcinoma of the head and neck (Beasley et al., 2002), gastric carcinomas (Kitadai et al., 2005) and pancreatic tumors (Rubbia-Brandt et al., 2004). In breast cancer, no correlation has been found between the number of lymphatic vessels and lymph node status, to the survival of patients (Bono et al., 2004). The multivariate analysis showed that high-density peritumoral lymphatic vessels were associated with higher risk of metastasis to lymph nodes in squamous cell carcinomas of the head and neck (Kyzas et al., 2005). In other studies, no correlation was found between the density of lymphatic vessels and lymphatic vessel invasion, lymph node status, and survival of patients as in cases of hepatocellular carcinoma and pancreatic cancer (Mouta Carreira et al., 2001; Sipos et al., 2005). In prostate cancer, or in some case series of breast cancer, there was no opportunity to determine the presence of lymphatic vessels inside the tumor. A lot of work highlights significant changes in the lymph vessels - the proliferation, budding of new blood vessels and expansion in the vicinity of the tumor.

As in physiological conditions, vascular growth factors VEGF-C and-D, activating receptor VEGFR-3, play an essential role in this process. VEGF-C and-D exhibit lymphangiogenic functions through the stimulation of VEGFR-3. They are produced as pre-propeptides that undergo proteolytic processing in the extracellular matrix. Their mature forms exhibit a greater affinity for VEGFR-3, but can also bind VEGFR-2 and induce angiogenesis. Overexpression of VEGF-C and-D in experimental tumor models was accompanied by intensive growth of new lymphatic vessels (Skobe et al. 2001), but in human tumors, these molecules involved in angiogenesis and lymphangiogenesis, are a controversial subject. Some authors state a significant correlation between expression of VEGF-C and-D and lymphangiogenesis and lymph node status, as well as an increase in the density of blood vessels in tumors (Mohammed et al., 2007; Nakamura et al., 2003; Nakamura et al., 2003) although there are also works in which this relationship is not stated (Currie et al., 2004).

In the mouse model of VEGF-C and VEGF-D secreted by tumor cells, there is induced formation of lymphatic vessels in, and around the tumor, which promotes the development of metastasis to regional lymph nodes. These processes have undergone deceleration under the influence of antibodies against VEGFR-3, blocking the activity of the ligands VEGF-D and C (Kitadai et al., 2005). Expression of VEGF-C was observed in many human cancers: breast cancer, cervical and bronchial and prostate and stomach cancers (Roskoski, 2007).

In a study of a small number of patients, including 17 cases of VIN and 26 cases of vulvar cancer, MacLean and colleagues demonstrated the presence of VEGF in 96% of vulvar cancer and only 6% of cases of VIN, not expressing this factor in healthy tissue (MacLean et al., 2000). In experimental models, tumor cells exhibiting overexpression of VEGF-C induced the formation of lymphatic vessels around the tumor (Saharinen et al., 2004; He et al., 2004). An additional issue is the relationship of lymphangiogenesis parameters such as density of lymph vessels, VEGF-C and D, and invasion of lymphatic vessels in tumor progression and prognosis.

In numerous trials, there has been a positive correlation between expression of VEGF-C and the invasion of lymphatic vessels, the presence of lymph node metastases and survival (He et al., 2004). Increased expression of VEGF-D was observed in breast, colorectal, gastric and thyroid multiforme and gliomas, and it has a positive correlation with the presence of lymph node metastases as reported in colorectal cancer, ovarian and bronchus (Roskoski, 2007). In malignant melanoma, VEGF-D likely has an important role both in lymphocytes and angiogenesis (Achen et al., 2001). Similarly, in many tumors, there has been a significant positive correlation between expression of VEGF- C, D in the primary tumor and lymph node status. However, in highly differentiated gastric cancers, this association is not found, and in cases of breast cancer and small cell lung cancer the results were inconclusive.

Increased expression of VEGF-C is a negative prognostic factor in many types of cancer, with the exception of Neuroblastoma, pancreatic cancer and colon cancer. A statistically significant correlation between VEGF-D expression in tumors and shorter overall survival was observed in endometrial cancer, ovarian and pancreatic cancers, in contrast to breast cancer or colorectal cancer (Thiele & Sleeman, 2006). Expression of VEGF-C in tumor cells in many types of cancer generally increases the risk of spread to the lymph nodes, and has some negative effect on survival (Hirakawa et al., 2007).

In gynecological tumors, one also considers the characteristics of lymphangiogenesis. In cervical cancer, a higher density of lymphatic vessels is observed in the periphery of the tumor and in the interior of the tumor in comparison with the normal cervix. The density of blood vessels in the periphery of the tumor correlates positively with higher tumor stage, lymphatic vessel invasion and metastases to lymph nodes, also an independent prognostic factor in multi-and one-dimensional analysis (Gombos et al., 2005). New development of lymphatic vessels has already been concluded in the early stages of carcinogenesis in cervical cancer. Longatto-Filho and colleagues observed that higher density of lymphatic vessels was characterized by changes in invasive cancers (squamous cell carcinoma and adenocarcinoma) compared with changes in preinvasive cancers (CIN I, CIN II, CIN III). However, not found in this study, was a statistically significant correlation between lymphatic vessel density and lymph node status (longatto-Filho et al., 2007). A widely used marker of lymphatic vessels is the D2-40. The expression of this marker was also found in cancer cells and cervical epithelium with features of CIN. Although D2-40 expression in the epithelium did not correlate with clinical features and histological changes, a significant relationship between low expression of D2-40 with invasion of lymphatic vessels and metastases in lymph nodes was observed. This may suggest participation of M2A antigen recognized by D2-40 in the interaction of tumor cells and endothelial cells of lymphatic vessels (Dumoff et al., 2005).

Similarly, expression of vascular growth factors essential for lymphangiogenesis in tumor cells is a phenomenon often described in cases of cervical cancer (Ueda et al., 2001). Increased expression of VEGF-C is attributed to the formation of lymph node metastases (Hashimoto et al., 2001).

A clear and statistically significant difference in the expression of VEGF-C and-D and their receptor VEGFR-3 was observed between (changes in) CIN I and CIN II, CIN III and invasive cancer. A higher degree of dysplasia was associated with increased expression of growth factors and their receptors. This may suggest, in addition to pro-lymphangiogenic activity, autocrine effects of VEGF-C and-D directly on tumor cells via receptor VEGFR-3 (Van Trappen et al., 2003). In cases of vulvar cancer and VIN-type changes described, we see the adverse effect that lymphangiogenic factors have on prognosis and their relationship with the progression of dysplastic lesions (Näyhä & Stenbäck, 2007; Lewy-Trenda et al., 2005). For many years, it was believed that the true VIN3 precursor of invasive vulvar cancer, created a higher risk for women over 40 years of age. Time from VIN3 diagnosis to invasive cancer is estimated to be about 4 years (1.1 to 7.3).

Recently, dominated by many views and defended by clinical data, is the belief that the potential for malignant changes in low grade VIN does not fully reflect the true behavior of these changes, but rather the survival of patients with this disease. (Rotmensch & Yamada, 2003; Van Seters et al., 2005; Jones et al., 2005). However, there are few reports analyzing the parameters of lymphangiogenesis in these diseases. In the present study, the presence of lymphatic vessels was observed along with changes in the type and degree of dysplasia. It was also found that expression of VEGF-C, VEGF-D and VEGFR-3 in both groups was correlated with the progression of the disease. No statistically significant difference between groups is most likely related to the small size of the analyzed series.

Taking into account the fact that lymph node involvement is an important prognostic factor in vulvar cancer (Raspagliesi et al., 2006), also having been described in other tumor lymphangiogenesis parameters, there is a strong gynecological association with progression and prognosis (Bednarek et al., 2008; Bednarek et al., 2009). One would expect that with vulvar cancer precursors and changes toward progressive disease, it is possible to determine similar relationships. Therefore, also bearing in mind the small number of publications on this matter, it seems that further analysis of lymphangiogenesis in the present group of cases of vulvar cancer is justified and purposeful. Due to the relatively small number of studies that have examined biomarkers (VEGF-C, VEGF-D and VEGFR-3) in carcinogenesis of the vulvar area and the lack of multivariate analysis in these studies, and taking into account the presence of small vulvar cancer and its precursors, one can not establish clear conclusions regarding their prognostic value.

7. References

Achen, MG., Williams, RA., Minekus, MP., Thornton, GE., Stenvers, K. &Rogers, PA. (2001). Localization of vascular endothelial growth factor-D in malignant melanoma suggests a role in tumour angiogenesis. *Journal of Pathology*, Vol.193, No.2, (February 2001), pp. 147-154.

Beasley, NJ., Prevo, R., Banerji, S., Leek, RD., Moore, J. &Van Trappen, P. (2002). Intratumoral lymphangiogenesis and lymph node metastasis in head and neck cancer. *Cancer Research*, Vol.62, (March 2002), pp. 1315-1320.

Bednarek, W., Mazurek, M. & Ćwiklińska, A. (2009). Ekspresja wybranych markerów i modulatorów angiogenezy u chorych na raka jajnika w okresie przed-, około- i pomenopauzalnym. *Ginekologia Polska*, Vol.80, No.2, (February 2009), pp. 93-98.

Bednarek, W., Wertel, I. & Kotarski, J. (2008). Limfangiogeneza w guzach nowotworowych. *Ginekologia Polska*, Vol.79, pp. 625-629.

Bono. P., Wasenius, VM., Heikkilä, P., Lundin, J., Jackson, DG. & Joensuu, H. (2004). High LYVE-1-positive lymphatic vessel numbers are associated with poor outcome in breast cancer. *Clinal Cancer Research*, Vol.10, No.21, (2004), pp. 7144-7149.

Currie, MJ., Hanrahan, V., Gunningham, SP., Morrin, HR., Frampton, C. & Han, C. (2004). Expression of vascular endothelial growth factor D is associated with hypoxia inducible factor (HIF-1alpha) and the HIF-1alpha target gene DEC1, but not lymph node metastasis in primary human breast carcinomas. *Journal of Clinal Pathology*, Vol.57, No.8, (August 2004), pp. 829-834.

Donoghue, JF., Lederman, FL., Susil, BJ. & Rogers, PA. (2007). Lymphangiogenesisof normal endometrium and endometrial adenocarcinoma. *Human Reproduction.*, Vol.22, No.6, (June 2007), pp. 1705-13.

Dumoff, KL., Chu, C., Xu, X., Pasha, T., Zhang, PJ. & Acs, G. (2005). Low D2-40 immunoreactivity correlates with lymphatic invasion and nodal metastasis in early-stage squamous cell carcinoma of the uterine cervix. *Modern Pathology*, Vol.18, No.1, (Jabuary 2005), pp. 97-104.

Gombos, Z., Xu, X., Chu, CS. & Acs, G. (2005). Peritumoral lymphatic vessel density and vascular endothclial growth factor C expression in early-stage squamous cell carcinoma of the uterine cervix. *Clinal Cancer Research*, Vol.11, (December 2005), pp. 8364-71.

Hashimoto, I., Kodama, J., Seki, N., Hongo, A., Yoshinouchi, M., & Okuda, H. (2001). Vascular endothelial growth factor-C expression and its relationship to pelvic lymph node status in invasive cervical cancer. *British Journal of Cancer*, Vol.85, No.1, (July 2001), pp. 93-7.

He, Y., Karpanen, T. & Alitalo, K. (2004). Role of lymphangiogenic factors in tumor metastasis. *Biochimica and Biophysia Acta*, Vol.1654, No.1, (March 2004), pp. 3-12.

Hillen, F. & Griffioen, AW. (2007). Tumour vascularization: sprouting angiogenesis and beyond. *Cancer Metastasis Review*, Vol.26, No.3-4, (December 2007), pp. 489-502.

Hirakawa, S., Brown, LF., Kodama, S., Paavonen, K., Alitalo, K. & Detmar, M. (2007). VEGF-C-induced lymphangiogenesis in sentinel lymph nodes promotes tumor metastasis to distant sites. *Blood*, Vol.109, No.3, (February 2007), pp. 1010-1017.

Iversen, T. & Tretli, S. (1998). Intraepithelial and invasive squamous cell neoplasia of thevulva: trends in incidence, recurrence, and survival rate in Norway. *Obstetrics & Gynecology*, Vol.91, No.6, (June 1998), pp. 969-972.

Ji, RC. (2006). Lymphatic endothelial cells, tumor lymphangiogenesis and metastasis: New insights into intratumoral and peritumoral lymphatics. *Cancer Metastasis Review*, Vol.25, No.4, (December 2006), pp. 677-694.

Jones, RW. & Rowan, DM. (1994). Vulvar intraepithelial neoplasia III: a clinical study of the outcome in 113 cases with relation to the later development of invasive vulvar carcinoma. *Obstetrics & Gynecology*, Vol.84, No.5, (November 1994), pp. 741-745.

Jones, RW., Rowan, DM. & Stewart, AW. (2005). Vulvar intraepithelial neoplasia: aspects of the natural history and outcome in 405 women. *Obstetrics & Gynecology*, Vol.106, No.6, (December 2005), pp. 1319-26.

Joura, EA. (2002). Epidemiology, diagnosis and treatment of vulvar intraepithelial neoplasia. *Current Opinion in Obstetrics and Gynecology*, Vol.14, No.1, (February 2002), pp. 39-43.

Judson, PL., Habermann, EB., Baxter, NN., Durham, SB. & Virnig, BA. (2006). Trends in the incidence of invasive and in situ vulvar carcinoma. *Obstetrics & Gynecology*, Vol.107, No.5, (May 2006), pp. 1834-22.

Kitadai, Y., Kodama, M., Cho, S., Kuroda, T., Ochiumi, T. & Kimura, S. (2005). Quantitative analysis of lymphangiogenic markers for predicting metastasis of human gastric carcinoma to lymph nodes. *International Journal of Cancer*, Vol.115, No.3, (June 2005), pp. 388-392.

Kuhn, L., Sun, XW. &Wright, Jr. TC (1999). Human immunodefficiency virus infection and female Lower genital tract malignancy. *Current Opinion in Obstetrics and Gynecology*, Vol.11, No.1, (Febryary 1999), pp. 35-39.

Kyzas, PA., Geleff, S., Batistatou, A., Agnantis, NJ. & Stefanou, D. (2005). Evidence for lymphangiogenesis and its prognostic implications in head and neck squamous cell carcinoma. *Journal of Pathology*, Vol.206, No.2, (June 2005), pp. 170-177.

Lewy-Trenda, I., Wierzchniewska-ławska, A. & Papierz, W. (2005). Expression of vascular endothelial growth factor (VEGF) in vulvar squamous cancer and VIN. *Polish Journal of Pathology*, Vol.56, No.1, pp. 5-8.

Longatto-Filho, A., Pinheiro, C., Pereira, SM., Etlinger, D., Moreira, MA. & Jubé, LF. (2007). Lymphatic vessel density and epithelial D2-40 immunoreactivity in pre-invasive and invasive lesions of the uterine cervix. *Gynecologic Oncology*, Vol.107, No.1, (October 2007), pp. 45-51.

MacLean, AB., Reid, WM., Rolfe, KJ., Gammell, SJ., Pugh, HE. & Gatter, KC. (2000). Role of angiogenesis in benign,premalignant and malignant vulvar lesions. *Journal of Reproductive Medicine*, Vol.45, No.8, pp. 609-612.

Markowska, J. (2006). Oncologia ginekologiczna. Wydawnictwo Medyczne Urban&Partner. Wrocław, Poland.

Maula, SM., Luukkaa, M., Grénman, R., Jackson, D., Jalkanen, S. & Ristamäki, R. (2003). Intratumoral lymphatics are essentialfor the metastatic spread and prognosis in squamous cell carcinomas of the head and neck region. *Cancer Research*, Vol.63, No.8, (April 2003), pp. 1920-1926.

Mohammed, RA., Green, A., El-Shikh, S., Paish, EC., Ellis, IO. & Martin, SG. (2007). Prognostic significance of vascular endothelial cell growth factors -A, -C and -D in breast cancer and their relationship with angio- and lymphangiogenesis. *British Journal of Cancer*, Vol.96, (March 2007), pp. 1092-1100.

Mouta Carreira, C., Nasser, SM., di Tomaso, E., Padera, TP., Boucher, Y. & Tomarev, SI. (2001). LYVE-1 is not restricted to the lymph vessels: expression in normal liver blood sinusoids and down-regulation in human liver cancer and cirrhosis. *Cancer Research*, Vol.61, No.22, (November 2001), pp. 8079-8084.

Nadja, ET. & Detmar, M. (2006). Tumor and lymph node lymphangiogenesis – impact on cancer metastasis. *Journal of Leucocyte Biology*, Vol.80, No.4, (October 2006), pp. 691-6.

Nakamura, Y., Yasuoka, H., Tsujimoto, M., Yang, Q., Imabun, S. & Nakahara, M. (2003). Flt-4-positive vessel density correlates with vascular endothelial growth factor-d expression, nodal status, and prognosis in breast cancer. *Clinal Cancer Research*, Vol.9, No.14, (November 2003), pp. 5313-5317.

Nakamura, Y., Yasuoka, H., Tsujimoto, M., Yang, Q., Imabun, S. & Nakahara, M. (2003). Prognostic significance of vascular endothelial growth factor D in breast carcinoma with longterm follow-up. *Clinical Cancer Research*, Vol.9, No.2, (February 2003), pp. 716-721.

Nathanson, SD. (2003). Insights into the mechanisms of lymph node metastasis. *Cancer.* Vol.98, No.2, (July 2003), pp. 413-23.

Näyhä, VV. & Stenbäck, FG. (2007). Increased angiogenesis is associated with poor prognosis of squamous cell carcinoma of the vulva. *Acta Obstetrica et Gynecologica Skandinavica.* Vol.86, No.11, (October 2007), pp. 1392-7.

Nisato, RE., Tille, JC. & Pepper, MS. (2003). Lymphangiogenesis and tumor metastasis. *Thrombosis and Haemostasis*, Vol.90, No.4, (2003), pp. 591-7.

Padera, TP., Kadambi, A., di Tomaso, E., Carreira, CM., Brown, EB. & Boucher, Y. (2002). Lymphatic metastasis in the absence of functional intratumor lymphatics. *Science*, Vol.296, No.5574, (June 2002), pp. 1883-1886.

Preti, M., Ronco, G., Ghiringhello, B. & Micheletti, L. (2000). Recurrent squamous cell carcinoma of the vulva: clinicopathologic determinants identifying low risk patients. *Cancer*, Vol.88, No.8, (April 2000), pp. 1869-1876.

Raspagliesi, F., Hanozet, F., Ditto, A., Solima, E., Zanaboni, F. & Vecchione, F. (2006). Clinical and pathological prognostic factors in squamous cell carcinoma of the vulva. *Gynecological Oncology*, Vol.102, No.2, (August 2008), pp. 333-7.

Roskoski, R. (2007). Vascular endothelial growth factor (VEGF) signaling in tumor progression. *Critical Reviews in Oncology/Hematology*. Vol.62, No.3, (June 2007), pp. 179-213.

Rotmensch, J. &Yamad, SD. (2003). Neoplasms oft he Vulva and Vagina. In: Kufe DW, Pollock RE, Weichselbaum RR, Bast JrR, Gansler TSHJFea, editors. Holland- Frei Cancer Medicine. 6th ed. Hamilton, Ontario: B.C. Decker, Inc.

Rubbia-Brandt, L., Terris, B., Giostra, E., Dousset, B., Morel, P. & Pepper, MS. (2004). Lymphatic vessel density and vascular endothelial growth factor-C expression correlate with malignant behavior in human pancreatic endocrine tumors. *Clinical Cancer Research*, Vol.10, No.20. (October 2004), pp. 919-6928.

Saharinen, P., Tammela, T., Karkkainen, MJ., Alitalo, K. (2004).Lymphatic vasculature: development, molecular regulation and role in tumor metastasis and inflammation. *Trends in Immunology*, Vol.25, No.7, (July 2004), pp. 387-95.

Sipos, B., Kojima, M., Tiemann, K., Klapper, W., Kruse, ML. & Kalthoff, H. (2005). Lymphatic spread of ductal pancreatic adenocarcinoma is independent of lymphangiogenesis. *Journal of Pathology*, Vol.207, No.3, (November 2005), pp.: 301-312.

Skobe, M., Hawighorst, T., Jackson, DG., Prevo, R., Janes, L. & Velasco, P. (2001). Induction of tumor lymphangiogenesis by VEGF-C promotes breast cancer metastasis. *Nature Medicine*, Vol.7, No.2, (February 2001), pp. 192-198.

Stacker SA, Caesar C, Baldwin ME, Thornton GE, Williams RA, Prevo R et al (2001). VEGF-D promotes the metastatic spread of tumor cells via the lymphatics. *Nature Medicine*, Vol.7, No.2, (February 2001), pp. 186-191.

Sundar, SS. & Ganesan, TS. (2007). Role of lymphangiogenesis in cancer. *Journal of Clinical Oncology*, Vol.25, No.27, (September 2007), pp. 4298-4307.

Thiele, W. & Sleeman, JP. (2006). Tumor-induced lymphangiogenesis: a target for cancer therapy? *Journal of Biotechnology*, Vol.124, No.1, (June 2006), pp. 224-241.

Thuis, YN., Campion, M., Fox. H, & Hacker, NF. (2000).Contemporary experience with the management of vulvar intraepithelial neoplasia. *International Journal of Gynecological Cancer*, Vol.10, No.3, (May 2000), pp. 223-227.

Ueda, M., Terai, Y., Kumagai, K., Ueki, K., Yamaguchi, H. & Akise, D. (2001). Vascular endothelial growth factor C gene expression is closely related to invasion phenotype in gynecological tumor cells. *Gynecological Oncology*, Vol.82, No.1, (July 2001), pp. 162-6.

Wissmann, C. & Detmar, M. (2006). Pathways targeting tumor lymphangiogenesis. *Clinical Cancer Research*, Vol.12, No.23, (2006), pp. 6865-8.

Van Seters, M., Van Beurden, M. & de Craen, AJ. (2005). Is the asssumed natural history of vulvar intraepithelial neoplasia III based on enough evidence? A systemic review of 3322 published patients. *Gynecological Oncology*, Vol.97, No.2, (May 2007), pp. 645-51.

Van Trappen, PO., Steele, D., Lowe, DG., Baithun, S., Beasley, N. & Thiele, W. (2003). Expression of vascular endothelial growth factor (VEGF)-C and VEGF-D, and their receptor VEGFR-3, during different stages of cervical carcinogenesis. *Journal of Pathology*, Vol.201, No.4, (December 2003), pp. 544-54.

Permissions

The contributors of this book come from diverse backgrounds, making this book a truly international effort. This book will bring forth new frontiers with its revolutionizing research information and detailed analysis of the nascent developments around the world.

We would like to thank Supriya Srivastava, for lending her expertise to make the book truly unique. She has played a crucial role in the development of this book. Without her invaluable contribution this book wouldn't have been possible. She has made vital efforts to compile up to date information on the varied aspects of this subject to make this book a valuable addition to the collection of many professionals and students.

This book was conceptualized with the vision of imparting up-to-date information and advanced data in this field. To ensure the same, a matchless editorial board was set up. Every individual on the board went through rigorous rounds of assessment to prove their worth. After which they invested a large part of their time researching and compiling the most relevant data for our readers. Conferences and sessions were held from time to time between the editorial board and the contributing authors to present the data in the most comprehensible form. The editorial team has worked tirelessly to provide valuable and valid information to help people across the globe.

Every chapter published in this book has been scrutinized by our experts. Their significance has been extensively debated. The topics covered herein carry significant findings which will fuel the growth of the discipline. They may even be implemented as practical applications or may be referred to as a beginning point for another development. Chapters in this book were first published by InTech; hereby published with permission under the Creative Commons Attribution License or equivalent.

The editorial board has been involved in producing this book since its inception. They have spent rigorous hours researching and exploring the diverse topics which have resulted in the successful publishing of this book. They have passed on their knowledge of decades through this book. To expedite this challenging task, the publisher supported the team at every step. A small team of assistant editors was also appointed to further simplify the editing procedure and attain best results for the readers.

Our editorial team has been hand-picked from every corner of the world. Their multi-ethnicity adds dynamic inputs to the discussions which result in innovative outcomes. These outcomes are then further discussed with the researchers and contributors who give their valuable feedback and opinion regarding the same. The feedback is then collaborated with the researches and they are edited in a comprehensive manner to aid the understanding of the subject.

Apart from the editorial board, the designing team has also invested a significant amount of their time in understanding the subject and creating the most relevant covers. They scrutinized every image to scout for the most suitable representation of the subject and create an appropriate cover for the book.

The publishing team has been involved in this book since its early stages. They were actively engaged in every process, be it collecting the data, connecting with the contributors or procuring relevant information. The team has been an ardent support to the editorial, designing and production team. Their endless efforts to recruit the best for this project, has resulted in the accomplishment of this book. They are a veteran in the field of academics and their pool of knowledge is as vast as their experience in printing. Their expertise and guidance has proved useful at every step. Their uncompromising quality standards have made this book an exceptional effort. Their encouragement from time to time has been an inspiration for everyone.

The publisher and the editorial board hope that this book will prove to be a valuable piece of knowledge for researchers, students, practitioners and scholars across the globe.

List of Contributors

Raghad Samir
Department of Obstetrics and Gynecology, Falun Hospital, Falun, Sweden

Dan Hellberg
Center for Clinical Research, Falun and Department of Women´s and Children´s Health, Uppsala University, Uppsala, Sweden

Narges Izadi-Mood, Soheila Sarmadi and Kambiz Sotoudeh
Department of Pathology, Tehran University of Medical Sciences, Iran

Oguntayo Olanrewaju Adekunle
Department of Obstetrics and Gynaecology, Ahmadu Bello University Teaching Hospital, Zaria Kaduna State, Nigeria

Supriya Srivastava
Cancer Science Institute, National University Singapore, Singapore

Kirvis Torres-Poveda, Ana I. Burguete-García, Margarita Bahena-Román, Alfredo Lagu-nas-Martínez and Vicente Madrid-Marina
Division of Chronic Infection Diseases and Cancer, Centro de Investigación Sobre Enfermedades Infecciosas, Instituto Nacional de Salud Pública, Cuernavaca, Morelos, Mexico

Chiung-Ru Lai, Chih-Yi Hsu and Anna Fen-Yau Li
Department of Pathology and Laboratory Medicine, Taipei Veterans General Hospital, Taiwan
School of Medicine, National Yang-Ming University, Taipei, Taiwan

Ana Ovanin-Rakić
Department of Gynecologic Cytology, Department of Gynecologic and Perinatal Pathology, Zagreb, Croatia

Isabelle Bourgault-Villada
AP-HP, Hôpital Ambroise Paré, Boulogne Billancourt, UVSQ, Versailles, France

Robert Jach, Małgorzata Radoń-Pokracka, Paulina Przybylska, Marcin Mika, Krzysztof Zając, Hubert Huras and Olivia Dziadek
Department of Obstetrics and Gynecology, Jagiellonian University Medical College, Kraków, Poland

Grzegorz Dyduch
Department of Pathology, Jagiellonian University Medical College, Kraków, Poland

Joanna Dulińska-Litewka
Department of Medical Biochemistry Jagiellonian University Medical College, Kraków, Poland

Joanna Streb
Department of Oncology, Jagiellonian University Medical College, Kraków, Poland

Klaudia Stangel-Wojcikiewicz
Department of Gynecology and Oncology, Jagiellonian University Medical College, Krakow, Poland

Printed in the USA
CPSIA information can be obtained
at www.ICGtesting.com
JSHW011414221024
72173JS00004B/546